# Our Knowledge Is Not Primitive

*The Iroquois and Their Neighbors*
Christopher Vecsey, *Series Editor*

OTHER BOOKS IN THE IROQUOIS AND THEIR NEIGHBORS SERIES

*Archaeology of the Iroquois: Selected Readings and Research Sources*
JORDAN E. KERBER, ed.

*Big Medicine from Six Nations*
TED WILLIAMS; DEBRA ROBERTS, ed.

*The Collected Speeches of Sagoyewatha, or Red Jacket*
GRANVILLE GANTER, ed.

*The History and Culture of Iroquois Diplomacy: An Interdisciplinary Guide to the Treaties of the Six Nations and Their League*
FRANCIS JENNINGS, ed.

*In Mohawk Country: Early Narratives about a Native People*
DEAN R. SNOW, CHARLES T. GEHRING, and WILLIAM A. STARNA, eds.

*Iroquoia: The Development about a Native World*
WILLIAM ENGELBRECHT

*Iroquois Medical Botany*
JAMES W. HERRICK; DEAN R. SNOW, ed.

*The Mashpee Indians: Tribe on Trial*
JACK CAMPISI

*Oneida Iroquois Folklore, Myth, and History: New York Oral Narrative from the Notes of H. E. Allen and Others*
ANTHONY WONDERLEY

*The Reservation*
TED C. WILLIAMS

*Seven Generations of Iroquois Leadership: The Six Nations since 1800*
LAURENCE M. HAUPTMAN

# Our Knowledge Is Not Primitive

## Decolonizing Botanical Anishinaabe Teachings

Wendy Makoons Geniusz

Illustrations by Annmarie Geniusz

Syracuse University Press

Syracuse, New York 13244-5290



First Paperback Edition 2023
23 24 25 26 27 28 6 5 4 3 2 1

Cover photograph by Annmarie Geniusz

∞The paper used in this publication meets the minimum requirements of the American National Standard for Information Sciences—Permanence of Paper for Printed Library Materials, ANSI Z39.48-1992.

For a listing of books published and distributed by Syracuse University Press, visit https://press.syr.edu.

ISBN: 978-0-8156-3204-7 (cloth)
978-0-8156-3806-3 (paperback)
978-0-8156-5652-4 (ebook)

**Library of Congress has cataloged the hardcover edition as follows:**

Geniusz, Wendy Djinn.
Our knowledge is not primitive: decolonizing botanical Anishinaabe teachings / Wendy Makoons Geniusz. — 1st ed.
p. cm. — (The Iroquois and their neighbors)
Includes bibliographical references and index.
ISBN 978-0-8156-3204-7 (hardcover : alk. paper)
1. Ojibwa Indians—Ethnobotany. 2. Ojibwa Indians—Ethnobotany—History—Sources. 3. Ojibwa Indians—Colonization. 4. Decolonization—United States. 5. Eurocentrism. 6. Ethnology—Social aspects—United States. 7. Ethnology—Research—United States. I. Title.
E99.C6G647 2009
325'.3973—dc22 2008055508

*Manufactured in the United States of America*

*To Nookomis, Nimishoomis, and Makwa*

*And to Mom and Dad*

WENDY MAKOONS GENIUSZ is professor of Indigeneity and Decolonization in the Department of Sociology at York University. She is of Métis and Cree descent and a proud Ojibwe language teacher.

# Contents

# Illustrations



TABLES

# Preface

I was raised with the anishinaabe teaching that one must always introduce the source of one's teachings. Before presenting this research, therefore, I introduce those elders who shared information with me, and without whose assistance this project would not have been possible.

The late Keewaydinoquay, a *mashkikiiwikwe* (medicine woman) and ethnobotanist, was one of my first teachers of *anishinaabe-gikendaasowin* (anishinaabe knowledge). She identified herself as *ajijaak* (Crane) Clan. She led Midewiwin ceremonies and trained *oshkaabewisag* (apprentices) to continue her work as a medicine woman and spiritual leader. Keewaydinoquay was born in 1918 (Tanner, pers. comm.).[1] She said that she spent much of her childhood in an anishinaabe village on Cat Head Bay, which is on the tip of the Leelanau Peninsula in Michigan (Tanner, pers. comm.; M. Geniusz, pers. comm.). At approximately nine years old, she was apprenticed to Nodjimahkwe, a well-respected mashkikiiwikwe in her village (Keewaydinoquay 1989a). As a child, she was one of only five children in her village who was not taken away to boarding school, giving her the opportunity to visit with and learn from all the elders in the village, many of whom greatly missed their own grandchildren away at school. She says that by the time she

1. This date comes from a letter that Keewaydinoquay's mother wrote as the date of Keewaydinoquay's birth in a letter she wrote to her own father, Keewaydinoquay's grandfather (Tanner, October 14, 2004). A copy of this letter is in the possession of Helen Hornbeck Tanner.

realized the great extent of knowledge that these elders and Nodjimahkwe had taught her, it was too late to thank them. She decided that sharing this knowledge with others would be the next best thing, and so she spent much of her life doing that (Keewaydinoquay 1991a). She founded the Miniss Kitigan Drum, a nonprofit organization dedicated to teaching anishinaabe culture, in one effort to preserve and teach this knowledge. Keewaydinoquay was also an ethnobotanist and taught courses, beginning in 1981, on philosophy and ethnobotany at the University of Wisconsin-Milwaukee (Keewaydinoquay n.d.*a*). She held several formal academic degrees, including a Bachelor of Science degree from Central Michigan University, which she received in 1944, and a Master of Science in Biology degree from Central Michigan University, which she received in 1977 (Keewaydinoquay n.d.*c*). She also completed the coursework for a Ph.D. in biology with an emphasis in ethnobotany at the University of Michigan, but she was unable to complete the required exams to begin writing her dissertation (W. Geniusz 2005, 193).

My mother, Mary Siisip Geniusz, was another of my first teachers of anishinaabe-gikendaasowin. Geniusz is Makwa (bear) Clan and was born in Cornwall, Ontario, in 1948. She is Cree, but she was one of Keewaydinoquay's oshkaabewisag and, as such, practices anishinaabe culture. Geniusz also worked as a teaching assistant for Keewaydinoquay's university courses. She did the majority of her work with Keewaydinoquay when I was between the ages of five and twelve. Many times during my childhood, my mother and I would collect botanical materials to make everything from mats for a wigwam to cough medicine. More recently, seeing this project as an opportunity to enrich my cultural and academic lives simultaneously, Geniusz has worked with me on even more ethnobotanical projects. Geniusz has a master's degree in liberal studies from the University of Wisconsin-Milwaukee, and she is completing a Master of Indigenous Knowledge degree from Seven Generations Education Institute in Ontario. Geniusz currently teaches courses on anishinaabe ethnobotany for Minnesota State University-

Moorhead's American Multicultural Studies Department and the University of Wisconsin-Milwaukee Continuing Education Program.

My husband's grandparents, George McGeshick, Sr., and his late wife, Mary McGeshick, also taught me many things about anishinaabe culture and language. Mary McGeshick had already passed over before I began this research, but many of the teachings that George McGeshick shared with me for this research came from experiences that he had had with his wife. George McGeshick was born in 1914. He is *Wawaazisii* (bullhead) Clan and a fluent Ojibwe speaker. McGeshick is chief of the Chicaugon Chippewa of Iron River, Michigan, and he has led that community for many years. He is also enrolled in the Mole Lake Band of Lake Superior Ojibwe. He remembers living with the Mole Lake Band when they lived at Pickerel Dam in Wisconsin before they were granted a reservation in the 1930s. McGeshick is a well-known birch bark canoe maker, who, with the help of his wife and family, has made canoes for many organizations, including the Smithsonian Institution.

Dora Dorothy Whipple, whose anishinaabe name is Mezinaashiikwe, has also taught me many things about anishinaabe-gikendaasowin and the Ojibwe language. Whipple is an elder from the Leech Lake Reservation in Minnesota, where she was raised near Boy River. Whipple's official birth date is November 9, 1919, but the records of her birth were lost in a fire at the Cass Lake office. Birth certificates were reissued after this fire, but Whipple says that those rewriting these records made guesses as to how old the people were. She thinks she was actually born in the fall of 1920. Whipple adds that although her birth certificate says she was born at Cass Lake, she was actually born five miles away (Whipple, pers. comm.). Today she is a respected member of Minneapolis's anishinaabe community. Whipple is a fluent Ojibwe speaker who has worked on many language revitalization projects, including the University of Minnesota's Ojibwe Language CD-ROM Project.

Ken Johnson, Sr., whose anishinaabe name is Waasebines, also worked with me on this research. Johnson is *Wazhashk* (muskrat) Clan, from the Seine River First Nation Reserve in Ontario. He is in his fifties and, unlike many of his generation, a fluent Ojibwe speaker. Johnson still practices the anishinaabe way of life and graciously shares his teachings with students from Canada and the United States. Johnson has worked on several Ojibwe language and culture revitalization projects, including the University of Minnesota's Ojibwe Language CD-ROM Project.

Another elder, who asked not to be identified, worked with me on this project. At her request, I identify her as Rose in this research. She wishes those reading this research to know only that she is from Canada.

## WRITING FROM AN ANISHINAABE PERSPECTIVE

Boozhoo. Mashkiigookwe indaaw. Makwa indoodem. Odinawemaaganan a'aw ninga onjibaawan iwidi Zhaaganaashiiwakiing gaye odinawemaaganan a'aw noos agaami-gichigamiing onjibaawan. Anishinaabewikwe aawi niiyawen'enyiban, Keewaydinoquay izhinikaazoban. Wiin Minis-gitigaaning onjibaaban. Minowakiing niin nindoonjibaa. Makoons indizhinikaaz, anishinaabewinikaazoyaan.

Hello. I am Cree and a member of the Bear Clan. My mother's people come from Canada, and my father's people come from Poland. My namesake was Ojibwe,[2] and her name was Keewaydinoquay. She came from Garden Island, Michigan. I come from Milwaukee, Wisconsin, and I am called Makoons.

2. In Ojibwe, *niiyawe'enh* (my namesake) refers to a reciprocal relationship between the person doing the naming and the one who is named.

It is in accordance with anishinaabe protocol that I introduce myself this way. According to our customs, I must explain who I am, to whom I am connected, and where I come from so that those listening to me will know the origin of my teachings. To do otherwise would be disrespectful to the many people who have sacrificed their time and energies to teach me these things. Genetically I carry a mix of Cree and non-native backgrounds. Culturally, however, I am anishinaabe because I was raised, in accordance with my Anishinaabe namesake's wishes, with the teachings of that culture. When writing this book, I am speaking from an anishinaabe perspective.

## ANISHINAABE NAMES

Anishinaabe names, commonly called Indian names, appear throughout this text. Out of respect for the individuals who have these names, no attempt has been made to analyze their meaning. In cases where I knew the individual, I let him or her choose how to spell his or her name. Otherwise, I chose the spelling of the name as found in the original source.

## TREATMENT OF OJIBWE WORDS

As will be explained in the introduction, Ojibwe words are an integral part of this text. Singular and plural forms of Ojibwe nouns are used throughout this text. The first occurrence of an Ojibwe word not quoted directly from a written document is followed by an English translation given in parentheses and cited in the glossary. There are no standard rules of capitalization for Ojibwe words, so I have chosen to capitalize only proper nouns. I have capitalized *Anishinaabe* only when referring to the Anishinaabe people, leaving the word in the lower case when using it to describe some facet of life or culture, such as "anishinaabe communities." I have used the diacritic ' to signify a glottal stop in Ojibwe words.

# Acknowledgments

I would like to thank all of the elders who helped me with this research. *Miigwech* (thank you) to: Keewaydinoquay, George and Mary McGeshick, Dora Dorothy Whipple, Ken Johnson, Sr., Rose, Cheryl Podgorski (Aukeequay) and Mary Geniusz. Without the knowledge you shared with me, I would not have been able to write this book.

For their support and suggestions throughout the years, miigwech to my thesis advisor, John D. Nichols, and my entire committee: Patricia Albers, Jean O'Brien-Kehoe, and David Martinez. Thank you to all the others who have given me invaluable support and suggestions: Helen Hornbeck Tanner, senior research fellow at the Newberry Library, Chicago; John Aubrey of the Newberry Library, Chicago; Douglas Harder of the Ronald E. McNair Program; and Robin Kornman and Diane Amour of the University of Wisconsin, Milwaukee. Thank you also to Neil T. Luebke of the Milwaukee Public Museum Botany Department.

I am grateful to all the people who helped fund my writing and research. Thank you to: Michigan State University for the Predoctoral Fellowship in American Indian Studies, the Newberry Library for the Frances C. Allen Fellowship, Colgate University for the John D. and Catherine T. MacArthur Predoctoral Fellowship in Native American and Educational Studies, and the American Studies Program at the University of Minnesota for the Graduate Summer Research Grant.

Miigwech to all my relations for their support and encouragement, without which I could not have completed this work. Thank you especially to: my parents, Robert and Mary Geniusz; my sister, Annmarie Geniusz, and her husband, Stephen Bockhold; my uncle, Edward Geniusz; my grandparents;

Lynn Ningwiisiisis Simonsen; and my clan sister, Mary Jane Allen, and her husband, Harvey Allen. Finally, miigwech to my husband, Errol Geniusz, who put aside his life so that he could support me in mine. Thank you for inspiring me to want to succeed and for encouraging me to do so.

# Our Knowledge Is Not Primitive

# Introduction

## *Decolonization and Biskaabiiyang Methodologies*

Many of us have this image, ingrained in our heads since primary school, of the colonists, those brave individuals of long ago who came to a "New World" and managed to make a life for themselves out of the bare wilderness. Ask any American child today what a colonist looked like, and she or he will probably describe men in black hats with huge silver shoe buckles and women wearing plain long dresses and shawls. We associate such images with the past, which is where we mentally place colonialism. For many adults, colonialism exists only in past centuries. After all, there were revolutions in the Americas; colonies broke away from their "mother" countries in Europe. Still, to some of us, and I speak here as a native person and a scholar in American Indian Studies, colonialism goes much further and much deeper in our society than these images, and it continues to be a driving force.

Linda Tuhiwai Smith, Maori scholar and author of *Decolonizing Methodologies,* argues that the colonizers and the colonized have viewed imperialism and colonialism in very different ways. From the perspective of the colonizers, imperialism was a state of mind that allowed Europeans to develop a sense of who they were and enabled them to "imagine the possibility that new worlds, new wealth and new possessions existed that could be discovered and controlled." European colonialism facilitated imperialism by "securing," "subjugating," and "exploiting" indigenous peoples, allowing imperialism to expand economically and to maintain control over new territories. Smith describes European colonies as "outposts" of imperialism, "cultural sites which preserved an image or represented an image of what the West or 'civilization' stood for." This image of civilization stood in sharp contrast to that of the indigenous peoples surrounding these outposts of imperialism (L. Smith 1999, 20–25).

Those who have been colonized often have an entirely different view of imperialism. From this perspective, colonialism is about one people completely taking over another people. It is not just about land. It does not end when one government gains control of another. It is about one society absorbing another society, and it continues until that process is accomplished. Yes, lands and governments are taken over, but so is every other facet of life, including language, culture, religion, knowledge, bodies, and beings. Many indigenous people, especially those trying to interpret imperialism and colonialism in order to understand what was done to them and their societies, look beyond the immediately noticeable results of imperialism, such as the colonization of land and the exploitation of peoples and resources, to the not so readily noticed results: the colonization of oneself. Linda Smith explains, "The reach of imperialism into 'our heads' challenges those who belong to colonized communities to understand how this occurred, partly because we perceive a need to decolonize our minds, to recover ourselves, to claim a space in which to develop a sense of authentic humanity" (1999, 21–23).

In the Americas, the early colonists used a variety of tactics to maintain and increase their landholdings. This part of the story usually is not debated. Yes, the history books tell us of wars between the colonists and the Indians. Yes, some colonists used underhanded tactics to attain land. Yes, some colonists also purchased land from the Indians, as exemplified by the existence of deeds, treaties, and other legal documents. For indigenous people, colonization was not just economic and physical exploitation and subjugation. It was also the exploitation and subjugation of our knowledge, our minds, and our very beings. This process was an important mechanism, which members of the colonial ruling elite used to gain control of the land and people in the "New World." To achieve and maintain their position of dominance over the land and its original inhabitants, members of the colonial ruling elite set into motion certain psychological, social, and economic mechanisms from which their descendants continue to benefit and because of which the majority of American Indians, other peoples of color, and the poor continue to suffer.

Over the last few decades scholars such as Howard Zinn, Ronald Takaki, George Lipsitz, and Elizabeth and Stuart Ewen have added to a growing body of literature on how the European ruling elite, at the beginning of American colonization, created images of race and the "other" in order to establish dominance in the "New World."[1] One multifaceted mechanism, which continues to maintain this power structure, is the colonization of knowledge. Those charged with carrying out various assimilation tactics were taught to view native knowledge as "primitive" or "evil," and, as a result, they often prevented its continued dispersal within native communities. Native people were also made to view their knowledge as "wrong" or "inferior" and non-native knowledge as "right" or "superior," and, having such views, many naturally chose what was made to look like the better knowledge.

The colonization of native knowledge assisted the colonizers in assimilating native peoples, but it also gave them another important benefit: They gained this knowledge for themselves. When looking at the colonization of botanical knowledge, the subject of this text, one sees that the colonizers did indeed gain much knowledge. In *Science and Colonial Expansion: The Role of the British Royal Botanic Gardens,* Lucile H. Brockway argues that not only did colonizers benefit from native botanical knowledge, they also were able to use this knowledge to fuel their imperialist efforts. Brockway does not describe the mechanisms of colonization that forced native peoples and others to view indigenous knowledge as primitive, but her examples illustrate the tremendous benefit the elite colonizers gained by making native knowledge appear "primitive." Once native people came to view their knowledge as inferior, some were willing to part with it, for a price reflecting its primitive, inferior nature. Others, seeing the devastating effects of assimilation efforts, chose to entrust this knowledge to researchers as a means of preserving it. In the end, the colonization process both destroyed and preserved native knowledge, and that is the beginning of this text.

This book examines the colonization of botanical *anishinaabe-gikendaasowin* (anishinaabe knowledge) and suggests ways that this information can be decolonized, reclaimed, and made useful to programs revitalizing anishinaabe language and culture. *Anishinaabe,* or *Anishinaabeg* in the plural, is the self-designation of several American Indian peoples, including the Ojibwe, Ottawa, and Potawatomi. There are contemporary anishinaabe

1. Please see references for examples of works by these authors.

communities in several states and provinces, including Michigan, Wisconsin, Minnesota, North Dakota, Montana, Saskatchewan, Manitoba, Ontario, and Quebec. Although these tribes have similar cultures and languages, there are differences between them, and because of those differences one should not assume that everything written in this book about the Anishinaabeg applies to all of these groups. References made to anishinaabe culture and language in this book refer specifically to those of the American Indian people who are commonly referred to in English as the Chippewa, Ojibway, Ojibwa, or Ojibwe. The Anishinaabeg call their language *Anishinaabemowin* or *Ojibwemowin,* although in English it is often called *Ojibwe.* Tribal stories confirm linguistic and anthropologic research demonstrating that the Anishinaabeg are related linguistically and culturally to other Algonquian tribes including the Menominee, Meskwaki, and Cree.

Researchers have recorded a fair amount of information about how the Anishinaabeg work with plants and trees; however, much of this information has been colonized. In order to use this knowledge for cultural revitalization, it must be reworked and reinterpreted into a format that is appropriate and usable to *anishinaabe-izhitwaawin* (anishinaabe culture). One could argue that any published text is colonized because colonizers brought the publishing industry to North America, but such an argument focuses on the immediately noticeable results of imperialism and the colonization process: one people controlling another people's land, government, and resources. While this is an important part of colonization and should not be devalued or overlooked, I choose to look deeper into the process of colonization by focusing on the often unseen mechanisms through which the colonists continue to maintain their power. In this context, a "text" refers to the written documentation of this research: a book, article, or unpublished note. A "colonized text" fits either or both of the following definitions: it serves the interests of the colonizers and the processes of systemic racism and oppression, or it presents information according to the philosophies, cosmologies, and knowledge-keeping systems of the colonizers, which are alien to those of anishinaabe-izhitwaawin.

Some of these colonized texts are insulting because they make degrading statements about the Anishinaabeg and their knowledge. Excuses for such statements are often made: "This writer is merely a product of the times in which she or he was writing." "This is only the opinion of the few people involved in the writing and publishing of this research." That may be, but these statements support something much larger and much more powerful than those who wrote or produced these texts. By insulting the Anishinaabeg and their knowledge, these texts are contributing to the mechanisms of colonization, those same mechanisms that made native peoples, other peoples of color, and the poor appear "inferior" and other people "superior." Some of these colonized texts are nearly unusable, or in some instances dangerous, because the presentation of the information in them is so abbreviated that one could not actually use the botanical material in the way suggested. As discussed in chapter 1, the academic disciplines out of which these researchers were writing and publishing often dictated how they presented this information. Although it would be inappropriate to blame these researchers for following their own traditions and those of their disciplines when presenting this information, those traditions are the traditions of the colonizers. In fact, most of these colonized texts, even those that are not abbreviated, present this information according to the philosophies and theories of the colonizer. This presentation of information is alien to anishinaabe culture, so for purposes of using this information in an anishinaabe cultural context, it must be reworked.

Anishinaabe people and organizations often attempt to use colonized texts of anishinaabe-gikendaasowin in their language and culture revitalization programs, but these texts often fall short of being adequate tools of cultural revitalization. For example, I saw an Anishinaabe woman at an organized cultural program demonstrating to a group of children how to make dream catchers, an ornament often made from twigs bent into a circle with webbing in the center.[2] The woman giving this demonstration did not understand that the written source she read aloud to her participants and used to make handouts had been written by an Asian scholar who tried to have her text translated into Ojibwe so that it would appear more "authentic." I only know of this scholar's efforts because she approached a friend of mine to make this translation. This text contains gross inaccuracies,

2. Twigs often used to make dream catchers are taken from one of two trees: willow (*Salix* spp.) or red osier dogwood (*Cornus sericea* L.).

including references to spirits who are not part of anishinaabe cosmology. The woman who used this text in her demonstration was not a scholar and was apparently unaware of the questionable nature of the material in this text. She was just looking for a written source of material to use in her demonstration. I have also seen language programs use *Plants Used by the Great Lakes Ojibwa,* a text that will be discussed in greater detail in the succeeding chapters, as their only resource for botanical anishinaabe-gikendaasowin. The authors of this text do not give enough details to make many of their descriptions usable, and they do not cite the sources of their botanical information. Further, researchers who did not know the language gathered many of the Ojibwe names presented in this text, and many of their plant names are thus unpronounceable. Despite these problems, this text is used by many anishinaabe programs, probably by virtue of the fact that it is one of the only widely available texts on the subject. Some of our people see this text and, possibly owing to its size, assume that it contains lots of information about how we use plants and trees. When describing my research on anishinaabe botanical knowledge to Anishinaabe people involved in cultural revitalization, I am often told, "Well, have you seen *Plants Used by the Great Lakes Ojibwa?* All the information you could possibly want on plants and trees is right there." This is not an accurate statement, and new materials need to be written before our elders, many of whom still hold this information, are no longer here to help us.

Before colonized texts can be used in a program revitalizing anishinaabe culture, they need to be "decolonized"; that is, changes need to be made to these presentations of anishinaabe-gikendaasowin. Degrading comments need to be removed, and additions need to be made to abbreviated instructions so that they can be safely followed. The presentation of this information needs to be brought back into the context of anishinaabe worldview, rather than left in the non-native context in which it currently exists. For example, researchers have split anishinaabe-gikendaasowin into separate areas of study, such as folklore, religion, music, and botany. Although the Anishinaabeg have stories, religions, music, and botanical information, these are not the extremely specialized, narrowly defined categories of non-native scholarly work. Within an anishinaabe cultural context, one does not ignore information in one of these categories in order to concentrate exclusively on another.

This research joins decolonization efforts in American Indian studies and in indigenous communities around the globe today. Although taking many forms, decolonization movements are ultimately based on the premise that indigenous people's lives and knowledge systems around the globe have been colonized along with their lands and resources, and it is now time to reverse the colonization process. Individuals from alien cultures have taken our lands, brainwashed our children, stolen our dead, and tried to make every part of our bodies and minds just like theirs. The colonizers have also collected and critiqued our knowledge and cultures, resulting in volumes of written materials that attempt to explain and present us to the rest of the world. Decolonization efforts seek to reclaim all these aspects of indigenous lives.

By breaking away from the colonizers and regaining control of our bodies, minds, and lives, those of us involved in decolonization attempt to reverse the colonization process. In *The Wretched of the Earth,* Frantz Fanon depicts the decolonizing process as a struggle between opposing binaries: the colonizer and the colonized. He calls for colonized peoples to decolonize and free themselves from European states, institutions, and societies. Fanon writes, "let us not pay tribute to Europe by creating states, institutions and societies which draw their inspiration from her. Humanity is waiting for something other from us than such an imitation, which would be almost an obscene caricature" (1968, 255). Fanon argues that decolonization is more than just breaking away from Europe; it is also about breaking away from European ideas and the structure of European institutions. It is about creating something new. More recently, in *Decolonizing Methodologies,* Maori scholar Linda Tuhiwai Smith applies the decolonization concept to academic theories, research, and methodologies. She describes this process not as discarding research, but rather as "centering our concerns and world views and then coming to know and understand theory and research from our own perspectives and for our own purposes" (1999, 39). According to Smith, then, decolonization is about changing, rather than imitating, the European or Western concept of research so that it fits into and can be used in conjunction with indigenous cultures.

The theoretical foundation of this decolonization of anishinaabe-gikendaasowin is based on premises similar to that of *Kaupapa Maori,* approaches to research developed by the Maori people as a means of combating the dehumanizing ways in which they have been depicted through academic research and of showing other Maori that they can use research to their own benefit. Before the development of Kaupapa Maori, the tendency of many Maori was to completely discard all research because of their people's negative history with research and researchers. Maori researchers, educated in non-native institutions, were trained to be detached from their research and found it difficult to work among their people until those working with Kaupapa Maori began developing methodologies for conducting research in one's own community. Kaupapa Maori accepts that Maori people have their own priorities and questions, which research can help answer and which are different from those of nonindigenous academics because they come from an entirely different worldview (L. Smith 1999, 183–93).

Like the Maori, we as Anishinaabeg have our own priorities and questions when conducting research of anishinaabe-gikendaasowin. Although our culture and language continue to live, there are pieces of both that are lost or are in danger of being lost because they are no longer practiced or used. Knowledge about working with plants and trees to make everyday items, food, and simple remedies is one of these areas. Our priorities in recording or reclaiming this information differ from those of non-native researchers, who often view their research on us as: a preservation effort, a final attempt to save strands of a dying culture, a bringing of native knowledge to the rest of the world, or a means of gathering data to prove some academic theory. Instead, our priority is to revitalize this knowledge within our own lives so that it will be there for our children and grandchildren and their children and grandchildren.

Given that the decolonization of anishinaabe-gikendaasowin has objectives so different from those of standard academic research, it stands to reason that the theoretical foundation of this research should come from a different philosophy than that of standard academic research. The Maori have found this to be true for their own research, and the foundations of Kaupapa Maori approaches to research come from Maori language, culture, teachings, and philosophy (L. Smith 1999, 183–93). The Anishinaabeg have an effort similar to that of the Maori. In 2003, the newly formed Masters of Indigenous Knowledge/Philosophy Program of the Seven Generations Education Institute, which is nestled between Couchiching First Nation

Reserve and Fort Francis, Ontario, on Agency One Land, began developing *Biskaabiiyang* approaches to research that attempt to decolonize the Anishinaabeg and their knowledge. This book implements Biskaabiiyang research methodologies to decolonize anishinaabe-gikendaasowin.

The Biskaabiiyang approach to research was developed in courses for the Masters of Indigenous Knowledge/Philosophy Program when students and instructors asked Anishinaabe elders to describe in Ojibwe certain concepts associated with how an Anishinaabe person learns within anishinaabe culture. The elders involved in this project include Delbert Horton and Ann Wilson of the Rainy River First Nations, Tobasonakwut Kinew of the Onigaming First Nation, and Edward Benton-Benai of Lac Courte d'Oreilles. The Anishinaabe academics involved in this project helped the students correlate these anishinaabe cultural concepts with those of traditional academia. This project resulted in a list of terms in the Ojibwe language, describing the anishinaabe way of being, to be used by those conducting research on anishinaabe-gikendaasowin. Those following Biskaabiiyang approaches to research use these research terms. Mary Siisip Geniusz, a master's student in this program, provided the copy of this list of terms used in this book.

In the context of this methodology, the word *biskaabiiyang* means "returning to ourselves" ("Anishinaabe Wordlist" 2003). Laura Horton, director of the Post Secondary Education Program at Seven Generations, describes Biskaabiiyang research as a process through which Anishinaabe researchers evaluate how they personally have been affected by colonization, rid themselves of the emotional and psychological baggage they carry from this process, and then return to their ancestral traditions (Horton, pers. comm.). As far as the survival of Anishinaabe people and culture is concerned, this is the most crucial part of Biskaabiiyang research methodologies. Not only does this approach to research give Anishinaabe academics and communities a common ground on which to begin talking about research, it also gives all of us a means of coming to terms with, to quote Linda Smith again, "The reach of imperialism into 'our heads.'" For generations Anishinaabe and all native peoples have been bombarded by negative stereotypes about our cultures and ourselves. Some of us have been conditioned to view ourselves and our cultures through the lenses created by those stereotypes. When using Biskaabiiyang methodologies, an individual must recognize and deal with this negative kind of thinking before conducting research. This is the only way to return to the teachings of our ancestors, and it is the only way to conduct new research that will be beneficial to the continuation of anishinaabe-gikendaasowin and anishinaabe-izhitwaawin.

The foundations of Biskaabiiyang approaches to research are derived from the principles of ***anishinaabe-inaadiziwin*** (anishinaabe psychology and way of being). These principles are ***gaa-izhi-zhawendaagoziyang***: that which is given to us in a loving way (by the spirits). They have developed over generations and have resulted in a wealth of ***aadizookaan*** (traditional legends, ceremonies); ***dibaajimowin*** (teachings, ordinary stories, personal stories, histories); ***Anishinaabemowin*** (language as a way of life); and ***anishinaabe-izhitwaawin*** (anishinaabe culture, teachings, customs, history). Through Biskaabiiyang methodology, this research goes back to the principles of anishinaabe-inaadiziwin in order to decolonize or reclaim anishinaabe-gikendaasowin.

For indigenous people, an important step in decolonization is taking control of research. The Maori have begun taking control of their own research through naming the approaches to research used in Kaupapa Maori. Linda Smith explains, "This naming of research has provided a focus through which Maori people, as communities of the researched and as new communities of the researchers, have been able to engage in a dialogue about setting new directions for the priorities, policies, and practices of research for, by and with Maori" (1999, 183). Decolonizing research is more than just having indigenous people working on research projects in their own communities. It is also about naming the research process and theories used to conduct that research. Therefore, this decolonization of anishinaabe-gikendaasowin, as demonstrated in the previous paragraph, includes the use of terms from the Ojibwe language to describe certain facets of anishinaabe knowledge and way of life. Table 1 provides the terms with which the reader will need to be familiar as they appear many times throughout this text.

Within the text, these terms are not explicitly marked with the word ***anishinaabe***. Following the conventions of Ojibwe discourse, once the explicit modified forms of such words are introduced, further references use the unmodified form with the fuller meaning understood. For instance, in the following chapters, izhitwaawin is used in place of anishinaabe-izhitwaawin.

1. ANISHINAABEMOWIN TERMS USED FREQUENTLY IN THIS RESEARCH

| *Explicit Modified Form* | *Unmodified Form* | *Gloss* |
|---|---|---|
| anishinaabe-gikendaasowin | gikendaasowin | knowledge, information, and the synthesis of our personal teachings |
| anishinaabe-inaadiziwin | inaadiziwin | anishinaabe psychology, way of being |
| anishinaabe-izhitwaawin | izhitwaawin | anishinaabe culture, teachings, customs, history |
| aadizookaan (sing.)<br>aadizookaanan (pl.) | | traditional legends, ceremonies |
| dibaajimowin (sing.)<br>dibaajimowinan (pl.) | | teachings, ordinary stories, personal stories, histories |

Because they come from our language, these terms more closely describe certain aspects of our knowledge and way of life. For example, we are not just talking about "knowledge," we are talking about anishinaabe-gikendaasowin, our own specific knowledge, unique to the Anishinaabe people, which includes not just information but also the synthesis of our personal teachings. Anishinaabe-gikendaasowin has been around since before the birth of the first human being. Anishinaabe authors Edward Benton-Banai and Basil Johnston outline the additions that Nenabozho, son of the West Wind *Manidoo* (spirit) and an Anishinaabe woman, made to anishinaabe-gikendaasowin (Benton-Banai 1988; Johnston 1976; Johnston 1995). As one of the original contributors to anishinaabe-gikendaasowin, Nenabozho categorized and labeled all of the earth's flora, fauna, and geographical features. Included in the information Nenabozho brought to the Anishinaabeg is guarded anishinaabe-gikendaasowin, such as *Midewaajimowin,* the teachings of the Midewiwin, an anishinaabe religious organization. One only receives such knowledge upon completion of certain degrees of training, and this knowledge is not the subject of this research. The anishinaabe-gikendaasowin discussed in this text is knowledge that every Anishinaabe needs to know for survival and in order to participate in anishinaabe-inaadiziwin and anishinaabe-izhitwaawin.

Finally, throughout this text I will use the first person when describing information that comes from my own life and experiences. Use of the first person is an important difference between Biskaabiiyang and other research methodologies. Biskaabiiyang approaches to research begin with the Anishinaabe researcher, who must look at his or her own life and how he or she has been personally colonized in order to conduct research from the standpoint of anishinaabe-inaadiziwin. Rather than assuming an unbiased stance to research, a researcher using Biskaabiiyang approaches to research submerges him or herself within anishinaabe-inaadiziwin and anishinaabe-izhitwaawin, the very things that he or she is researching. From this position, the Anishinaabe researcher must acknowledge his or her personal connection to the research he or she is conducting because, as will be explained more fully in chapter 2, the protocols of anishinaabe-izhitwaawin require that one always explain his or her personal and intellectual background whenever he or she shares an aadizookaan or dibaajimowin, such as those presented in this research. To do otherwise takes credibility away from the information presented and insults those who gave that Anishinaabe those teachings.

# 1

# The Presentation of Botanical Anishinaabe-gikendaasowin in the Written Record

## INTRODUCTION

This chapter discusses the documentation of botanical gikendaasowin over the last 170 years.[1] For those looking for this information in the written record, this chapter provides an overview of what is available.[2] It provides sample entries from these texts to give readers an idea of the quality and quantity of information available and the degree to which these texts may be useful to a revitalization program. Many of these texts contain colonized information that if reworked and decolonized could be very useful to programs revitalizing izhitwaawin. Some of them contain language or statements that serve the interests of the colonizers and the process of systemic racism and oppression. Most of these texts present information according to the philosophies, cosmologies, and knowledge-keeping system of the colonizers, which are alien to those of izhitwaawin. Chapter 3 discusses specific passages from these texts, exemplifying how they represent "colonized"

1. This calculation is based on the earliest list of Ojibwe plant names that I have found, that found in part 2 of John Tanner's captivity narrative (James [1830] 1956). There have been earlier texts written on plant knowledge from various tribes, including many journals of expeditions, but this is the earliest list I have found that is identified specifically as a list of Ojibwe plants and trees.

2. The reference list includes citations for sources containing a large amount of botanical gikendaasowin. Not all of these sources are discussed fully in this chapter.

texts. This chapter focuses on the history of how botanical gikendaasowin has been colonized. It is important for those working with these texts to be familiar with the backgrounds of their authors and the conditions under which they were written because this understanding can help to explain the presentation of this information and can aid us in reworking and decolonizing it. Although short details about how the Anishinaabeg work with plants and trees can be found in works covering almost any topic having to do with the Anishinaabeg, this chapter focuses on those sources presenting large portions of this knowledge.

## A BRIEF HISTORY OF DOCUMENTATION OF BOTANICAL KNOWLEDGE

The colonized texts of gikendaasowin are part of a much older non-native tradition of gathering and recording botanical knowledge, traceable at least as far back as 4000 B.C. to the clay tablets of Sumer, which contain botanical recipes for treating ailments, prayers, and other spiritual information used to help with the healing process (Erichsen-Brown 1979, vi). This tradition continued in Classical times. One of Aristotle's students wrote the earliest known Greek herbal around 300 B.C. (Collins 2000, 31). Dioscorides, a Greek physician living around A.D. 50–68, wrote the earliest Western herbal from which a complete pharmacopoeia has survived,[3] and this herbal, containing warnings about poisonous plants, information about when to gather certain plants, botanical recipes, and a discussion of the economic potential of some exotic plants, became the basis for European herbals throughout the Middle Ages. When the Europeans reached the Americas, they had only recently begun collecting and recording new botanical information from their local areas. Before the Renaissance, European herbals were nothing more than embellishments of Dioscorides' text (Collins 2000, 32; Davis 1995, 40–41).

Upon reaching the Americas, the Europeans encountered native peoples who also had extensive traditions of gathering, recording, or otherwise remembering knowledge about plants. Learning of this knowledge, the

3. The original text by Dioscorides has not survived, although early manuscript copies of it do (Collins 2000, 299).

Europeans continued their own traditions of documenting and utilizing it. Some of them began writing information about indigenous plant use in their journals and expedition reports. For example, in the account of his second voyage, 1535–36, Jacques Cartier describes how the Laurentian Iroquois, living at what is now Montreal, cured his crew of scurvy using a tea made from a local tree (Erichsen-Brown 1979, ix).[4] Expedition and "exploration" literature such as this continued to be written for several centuries, giving varying degrees of detail as to the origin of this knowledge. For instance, when describing a plant identified as "winter green" in his *Travels Through the Interior Parts of North America in the Years 1766, 1767, 1768,* Jonathan Carver writes, "The Indians eat these berries, esteeming them very balsamic, and invigorating to the stomach. The people inhabiting the interior colonies steep both the sprigs and berries in beer, and use it as a diet drink for cleansing the blood from scorbutic disorders" (1781, 510–11). Unlike the example in Cartier's report, here the reader is left to wonder which group of "Indians" Carver describes. Given that many such passages exist in the expedition literature, it is difficult to find in these documents botanical information coming from any specific tribe.

It seems that those in Europe were eager to read about botanical information from the Americas. Nicholas Monardes, a Spanish doctor from Seville, created the first herbal of plants from the Americas in 1569, and it was so popular that it was translated into Latin, French, and English (Rohde 1922, 120–41). John Gerard includes several plants from the Americas, along with the first published depiction of the potato, in his herbal, one of the most famous English texts of that genre (Gerard 1597; Rohde 1922, 93, 104). Gerard mentions the usage of certain plants by native people such as in this entry for "Corn," which he also identifies as *"Frumentum Indicum,"* and "Turkie corne":

4. Erichsen-Brown suspects this tree was either the "spruce" or the "hemlock" (1979, ix). Keewaydinoquay teaches that the use of "conifers" to treat scurvy is an important teaching in gikendaasowin. She explains that the high vitamin C content in all of the conifers cures scurvy, adding that the varieties of spruce, those of the family *Picea* spp., have the highest content of vitamin C, which they release into hot water faster than any of the other conifers (Keewaydinoquay 1988; M. Geniusz, pers. comm.).

> Turky wheate doth nourish far lesse than either Wheate, Rie, Barly or Otes. . . . We have as yet no certain proofe or experience concerning the vertues of this kinde of Corne, although the barbarous Indians which know no better, are constrained to make a vertue of necessitie, and think it a good food; whereas we may easily judge that it nourisheth but little, and is of hard and evill digestion, more convenient food for swine than men. (1597, 77)

Like his contemporaries and generations of researchers to come, Gerard makes no distinction between different groups of "Indians." He also belittles their knowledge of using this plant, saying that they "know no better" than to enjoy eating corn, which they are forced to do out of necessity. As in the expedition literature, herbals are also generally unhelpful when trying to identify information from a specific tribe.

The search for botanical knowledge helped to drive the force of European colonization in the Americas and around the world. Many "explorers" and their sponsors sought the economic opportunities brought by "discovering" previously unknown plants and botanical knowledge. In 1494, for instance, Henry VII of England sent John Cabot to North America specifically to look for spices and medicines, which he intended to sell in his new spice and drug depot. He hoped his depot would be larger than that in Alexandria, the contemporary capital of that trade (Erichsen-Brown 1979, x). In *Science and Colonial Expansion,* Lucille Brockway argues that the search for botanical indigenous knowledge encouraged, fueled, and made possible European colonization of the globe. Initially, Europeans brought plants from the Americas back to Europe and Asia, improving nutritional quality and quantity of food, which caused a population explosion and created a physical force for colonization (1979, 39–46). Later, European research institutions, such as the Botanic Gardens at Kew, began experimenting with plants and indigenous botanical knowledge from the Americas, creating products to benefit British health and economy as well as those of their colonies. Medicinals from the Americas, including the malaria treatment found in cinchona bark, helped European colonists settle in other tropical regions of the world such as India (1979, 103–33). The colonizers also used indigenous knowledge to exploit the people from whom it came. In 1810, for example, a rubber industry began in Brazil, which used indigenous knowledge of gathering natural rubber while forcing indigenous people to do all of the hard labor involved in this process (1979, 144–50).

Although coming from different histories and locations, similar circumstances affected the documentation of gikendaasowin and indigenous knowledge in other parts of the Americas. Personal employment and prestige were certainly factors for these collectors. The prospect of being the first to "discover" some portion of indigenous botany was undoubtedly appealing to them. There was also the excitement of being the one person to document a people's remaining knowledge before it was gone forever. It is important to acknowledge the connections between the colonization of anishinaabe botanical knowledge and that of other native peoples because, essentially, the same forces and philosophies recorded and appropriated all of this knowledge. From this history, we find other native groups in similar states of colonization; together we can build decolonization strategies.

## OJIBWE WORDLISTS

The earliest texts dedicated specifically to the documentation of botanical gikendaasowin are lists of Ojibwe plant names. The vocabularies found in part 2 of the 1830 publication of *A Narrative of the Captivity and Adventures of John Tanner,* edited by Edwin James, which are often omitted from republications of this text, contain the oldest of these. One of the vocabularies, "Catalogue of Plants and Animals," includes lists of Ojibwe names for trees, plants, and animals, each with varying degrees of identification and usage information. Sources are not given for this botanical information. All the text says is that Edwin James edited parts 1 and 2, but John D. Nichols, who is editing these vocabularies, suspects that this material is from John Tanner (Nichols, pers. comm.). James and Tanner worked together on several Ojibwe texts, including translating the gospels together in 1830 and 1831 (Fierst 1986, 31). Both men also worked for the U.S. government. Tanner was an interpreter, James an army medical doctor (Fierst 1996, 227). The lists in this vocabulary are divided into sections, each having an Ojibwe and an English title, such as: "Ne-bish-un—Trees with broad leaves" (James [1830] 1956, 293, 297). As seen in this example, these categories appear to follow more closely a non-native scientific perspective than an anishinaabe one, especially given that *ne-bish-un,* or *aniibiishan,* literally means "leaves or tea" (Nichols and Nyholm 1995, 11). In some entries, little information is given about the plant listed, such as the following entry: "Sah-sah-way-suck—Turkey potatoes" (James [1830] 1956, 297). Here James gives little identification for this plant, other than putting it into the category of "Wea-gush-koan—Weeds, or herbaceous plants." In other entries, he gives more identification information, such as "Ba-se-kwunk—This is a red astringent root, much valued by the Indians, as an application to wounds. Avens root?" (James [1830] 1956, 299). Although this example is one of the few entries in which James does more than simply list the Ojibwe name of a plant, he also questions whether this plant is "Avens root."

The next published wordlist of Ojibwe plant names is found in the short grammar at the end of Andrew J. Blackbird's *History of the Ottawa and Chippewa Indians of Michigan: A Grammar of Their Language, and Personal and Family History of the Author,* published in 1887 (107–28). Unlike Tanner's captivity narrative, we know that Blackbird himself wrote this text because an editor's note at the end of this text reads, "This work is printed almost verbatim as written by the author" (Blackbird 1887, 128). Blackbird was educated in the non-native tradition, and he worked as a U.S. interpreter at Traverse City, Michigan, for many years (1887, 3). In his introduction, Blackbird says that he has included these vocabularies and the other language materials in his text for the interest of "all who may wish to inquire into our history and language" (1887, 5). He says that he wrote this text to correct the writings of previous authors who attempted to write about the Ottawa and Chippewa, all of which he has found to be inaccurate (1887, 7). In a section entitled "The Author's Reasons for Recording the History of His People, and Their Language," Blackbird writes,

> The Indian tribes are continually diminishing on the face of this continent. Some have already passed entirely out of existence and are forgotten, who once inhabited this part of the country. . . . My own race, once a very numerous, powerful and warlike tribe of Indians, who proudly trod upon this soil, is also near the end of existence. In a few more generations they will be so intermingled with the Caucasian race as to be hardly distinguished as descended from the Indian nations, and their language will be lost. (1887, 24)

Blackbird's grammar, then, is an attempt by an Ojibwe speaker to save some portion of his language from what he sees as inevitable extinction. The Ojibwe names for plants, trees, nuts, and fruit include the following: "Au-zhaw-way-mish, pl. eg beech tree" and "Wau-be-mis-kou-min, pl. og; white raspberry" (1887, 123). As seen in these examples, Blackbird identifies the trees and plants in his list by Ojibwe and common English names only. However, he gives names for only a few trees and plants, and most of the ones he does list, including the black ash and black raspberry, are so common that the names he uses may provide enough identification information.

At this point one might ask: how are these wordlists examples of colonized texts? This question is exacerbated by the identity of these authors. Clearly James is identified as editor of the earlier set of vocabularies, but if Tanner is indeed the source of the information in these lists, and if he had a hand in writing this text, as might be assumed given that part 1 of the text tells his life's story, then one might argue that this cannot be a colonized text because it is written by someone who lived much of his life as an Anishinaabe. This assumption is even truer of Blackbird, who identifies himself as Anishinaabe and who claims to be using this text as a means of preserving some of his people's history and language. Blackbird, Tanner, and James all worked for the U.S. Indian Service or the army. They were part of the system that attempted to subjugate and oppress the indigenous peoples of this country. Does that make their texts "colonized"? Well, not necessarily, but it is an important point to make because it shows the perspective out of which they were writing. Even if they held these positions purely for the economic benefits, they were a part of and could have been influenced by the colonizing structure. Ultimately, however, it is the presentation of information in these texts that makes them "colonized" and proves that James, Blackbird, and possibly Tanner, if he was involved, were influenced by the system of which they were a part. The practice of gathering lists of words or names is part of non-native knowledge-keeping systems, not those keeping gikendaasowin. While the practice of making vocabularies is not inherently wrong or detrimental to gikendaasowin, it is still part of an alien knowledge-keeping system. As explained in chapter 3, such wordlists are wonderful examples of texts that can be decolonized, reworked, and made useful to programs revitalizing izhitwaawin, but the language in them needs to be checked so that the words may be properly used and pronounced. If the goal of a program is to present botanical gikendaasowin within the context of izhitwaawin, then these wordlists will also need to be incorporated into a format that does more than simply list plant names.

## PUBLISHED ANTHROPOLOGICAL SOURCES

Shortly after the publication of Blackbird's list, anthropological sources began publishing botanical gikendaasowin. The Bureau of American Ethnology, or BAE, published the earliest. Originally known as the Bureau of Ethnology, the BAE was founded on March 3, 1879, when an act of Congress combined several existing government surveys into the U.S. Geological Survey and gave the anthropological research that had been carried out under those government surveys to the Smithsonian Institution. Major John Wesley Powell, who had been the director of one of the surveys combined under the Geological Survey, was given the responsibility of becoming the Bureau of Ethnology's first director (Powell 1881, xi–xii; Judd 1967, 3–6). The Smithsonian, and later the BAE, dominated American anthropology from the founding of that institution in 1846 until the death of BAE Director Powell in 1902 (Hinsley 1981, 9). These institutions were interested in getting information from anyone who had any contact with Indian people, and of course it must be remembered that, unlike today, this was the beginning of anthropology, when there was little, if any, professional training in ethnology. The first secretary of the Smithsonian, Professor Joseph Henry, sent questionnaires to army officers, missionaries, merchants, and others who had regular contact with Indian peoples, and he received lots of data, some more useful than others, in response. Powell began working with this data before and after becoming director of the BAE and used it to publish eight volumes entitled *Contributions to North American Ethnology* (Judd 1967, 7–9). As director of the BAE, Powell continued asking people who had contact with Indians to send the BAE information. In his introduction to the *Seventh Annual Report of the Bureau of Ethnology*, Powell writes,

> The present opportunity is used to invite and urge again the assistance of explorers, writers, and students, who are not and may not desire to be

> officially connected with this Bureau. Their contributions, whether in the shape of suggestion or of extended communications, will be gratefully acknowledged and carefully considered. If published in whole or in part, either in the series of reports or in the monographs or bulletins, as the liberality of Congress may in future allow, the contributors will always receive proper credit. (1891, xv–xvi)

The BAE apparently received many papers in response to such advertisements, giving them ample materials to publish. Until it stopped being a separate bureau of the Smithsonian and merged with the U.S. National Museum's Department of Anthropology on July 29, 1964, to form the Smithsonian's Office of Anthropology, the BAE published eighty-one annual reports. The BAE also published 193 bulletins, beginning in 1886 and ending in 1964 (Judd 1967, 7). The bulletin series began when Director Powell saw that the Bureau was receiving many papers, which, as he wrote to Secretary Baird of the Smithsonian, "should be published but are scarcely worthy of publication in the Annual or in the quarto series" (as cited in Judd 1967, 96–97). When Congress gave the BAE funding in 1886 and 1888, the bulletin series began. After fiscal year 1931, the BAE reports no longer included accompanying ethnographic essays, and these papers were then published in the bulletin series (Judd 1967, vii, 78–79, 112).

While the BAE was still a fairly new organization, other anthropological organizations began incorporating themselves and producing their own publications. Among these were the American Anthropological Association, publisher of the *American Anthropologist,*[5] and the Wisconsin Archeological Society, publisher of the *Wisconsin Archeologist.* The American Anthropological Association, incorporated in 1902, was created as an attempt to unite American anthropologists into a national organization. The purpose of this organization, as stated in its articles of incorporation, is "to promote the science of anthropology, to stimulate and coordinate the efforts of American

5. The Anthropological Society of Washington had a journal called the *American Anthropologist,* but when moves were made to start a national anthropological organization, this society discontinued their journal and gave its name to the national organization (AAA 1903, 179).

anthropologists, to foster local and other societies devoted to anthropology, to serve as a bond of union among American anthropologists and anthropologic organizations present and prospective, and to publish and encourage the publication of matter pertaining to anthropology" (American Anthropological Association [AAA] 1903, 178, 182–82). Early officers of this organization include prominent names in anthropology of the time, including W. H. Holmes, F. W. Hodge, and Major Powell (AAA 1903, 190).[6] Articles from the *American Anthropologist* are some of the most readily available published sources on gikendaasowin as this journal can be read and printed from the online database JSTOR. The Wisconsin Archeological Society was incorporated in 1903, although the first volume of their journal was published in 1901. Articles found in the *Wisconsin Archeologist* relate to the archeology of Wisconsin and surrounding states (Wisconsin Archeological Society [WAS] 2003). Like authors of papers published by the BAE, authors of articles published by the AAA and the WAS during the late nineteenth and early twentieth centuries included any individuals who had any information to share about native peoples. Both the AAA and the WAS are still very active and continue to produce the journals mentioned here, as well as other publications.

The BAE, and arguably its contemporary anthropological organizations given that they were also staffed and connected with those running the BAE, was an integral part of the mechanisms that supported the colonizing structures in North America. It was part of the federal government, and many of those working and researching for it were also personally connected to the subjugation and oppression of Indian peoples. The BAE's first director, Powell, had been a major of artillery in the Union Army during the Civil War (Judd 1967, 4). Although this position did not necessarily put him in armed engagements with Indian peoples, Powell still fought for the same colonizing force that did.

Those submitting research to BAE publications were often themselves working to colonize Indian peoples because these were the people who, in their jobs as teachers, members of the military, and Indian agents, came into contact with and could gather information from Indian peoples. As with the

6. It appears that Powell died before he could take his position as vice president in 1903 (AAA 1903, 190).

wordlists, the background of their authors does not condemn the information in these texts, nor does it automatically make them "colonized," but it does give us a perspective out of which these texts were recorded and written, and it does suggest certain biases.

The first anthropological publication containing a substantial amount of botanical gikendaasowin was "The Midē'wiwin, or 'Grand Medicine Society' of the Ojibwa," written by Walter James Hoffman and published by the BAE. Hoffman was a medical doctor who had worked as an army surgeon. In 1877 he was put in charge of the ethnological and mineralogical collections of the U.S. Geological Survey. When the BAE was founded, Hoffman became an assistant ethnologist for that bureau and conducted fieldwork among Indian tribes across the United States and Canada (Chamberlain 1900, 44). Hoffman worked with the Ojibwe long enough to publish several ethnological articles on that tribe.[7]

"The Midē'wiwin" includes six pages listing plants identified as those that candidates for the Midewiwin learn at various times in their training (Hoffman 1891, 197–201, 226). Here is a sample entry:

> *Quercus alba*, L. White Oak. Mītig'ōmish'.
> 1. The bark of the root and the inner bark scraped from the trunk is boiled and the decoction used internally for diarrhea.
> 2. Acorns eaten raw by children, and boiled or dried by adults. (1891, 198)

In this example, as in most of his entries, Hoffman identifies the plant by scientific, common, and in most cases an Ojibwe name. For some Ojibwe names he also gives a literal translation. This is the approximate length of most of his entries. As seen here, Hoffman gives very little instruction as to how to gather, prepare, or use any of the recipes described in this list. For example, he says that a "decoction" of the inner bark of this tree is "used internally for diarrhea," but he does not say how strong a decoction should be used or how often it should be administered (1891, 201–2).

Frances Densmore, whose research appears in BAE reports and bulletins and in *American Anthropologist* articles, was the second person to publish

7. For examples see Hoffman 1891; Hoffman 1889; and Hoffman 1888.

anthropological texts on botanical gikendaasowin, the two most detailed being "Uses of Plants by the Chippewa Indians" and *Chippewa Customs* ([1928] 1974; [1929]1979). Densmore was a musician who worked closely with a number of Indian tribes because of her interest in recording Indian songs, a project that resulted in approximately 2,500 wax cylinder recordings (Vennum 1980, 185). Considering her background as a musician, one might be surprised that Densmore took the time to research and write texts on other aspects of anishinaabe culture. She explains that while collecting songs for *Chippewa Music* (1910, 1913), Anishinaabe consultants sang her healing songs, offering to show her the plants connected to those songs and to explain how they are used ([1928] 1974, 281). She later says that studying anishinaabe songs "led to a friendliness with the people and a willingness on their part to give information concerning their customs" ([1929] 1979, 1).

Densmore was one of the BAE's few paid collaborators, but not a member of its staff. Her relationship with the BAE began after she conducted some of her first field investigations with the Anishinaabeg in 1907 and asked the BAE for financial support to study and record American Indian customs, which she viewed as being on the verge of extinction. They agreed to assist her (Densmore 1941, 528–29; Archabal 1977, 101).

Other anthropological organizations became interested in her research, and when the BAE brought her to Washington in 1907 to present her research,[8] the American Anthropological Society invited her to give a lecture for them. They repeated this invitation when she came to the BAE to report on the completion of her Chippewa music research (Densmore 1941, 529; Archabal 1979, 102). Densmore also published research in this society's journal, *American Anthropologist,* including a brief account, "An Ojibwa Prayer Ceremony," in their "Anthropologic Miscellanea" section and a transcript of a speech she gave at the Central States Branch entitled "The Native Art of the Chippewa" (Densmore 1907, 443–44; Densmore 1941, 678–81).

From her own description of her research methods, it is clear that Densmore sees her research as a preservation tool for future generations, including

8. Densmore says that this was in 1907, but Archabal says it was in 1908 (Densmore 1941, 529; Archabal 1977, 102).

future generations of native peoples. She also claims to have self-imposed restrictions on what kinds of information she is willing to record. After seeing how her "friend" Maingans[9] was banned from continuing to attend Midewiwin ceremonies after allowing her to record his Midewiwin songs, Densmore says that she stopped accepting "material that is surrounded by superstition." She explains, "I tell the Indians that I am trying to preserve the material so their children will understand the old customs, and that I do not want them to worry or be unhappy after I have gone. Sometimes this allays their fears and they are willing to talk freely, but I would rather miss some information than cause such distress as that of my old friend, Maingans, the Chippewa" (1941, 529–30). Later she writes of her work with Indian music, saying, "My work has been to preserve the past, record observations in the present, and open the way for the work of others in the future" (1941, 550).

Densmore spent approximately twenty years collecting information from the Anishinaabeg ([1929] 1979, 1), and both "Uses of Plants by the Chippewa Indians" and *Chippewa Customs* contain much gikendaasowin as a result. Densmore divides the information she presents in "Uses of Plants by the Chippewa Indians" into categories describing how plants are used as food, medicine, dyes, "charms," and "useful and decorative arts." Perhaps acknowledging the limitations of her research abilities, she had scientists working for various government institutions identify her plant specimens, classify the diseases for which they are used, and describe their medicinal properties and medical constituents ([1928] 1974, 281). Densmore provides descriptions of varying lengths of how the Anishinaabeg use certain plants and trees. Her "Plants as Medicine" section consists of some detailed descriptions of how plant knowledge is maintained within anishinaabe culture, as well as some descriptions of various medical procedures. She presents the medical recipes in this section in a chart. The abbreviated nature of the recipes presented in this chart, coupled with the fact that it is printed on separate facing pages, makes this chart hard to decipher. In one of these, Densmore gives a recipe to make a treatment for a cold using a plant identified as "*Acorus calamus* L. (Calamus)." Indicating that the root of this plant is used, Densmore provides the following instructions on how to prepare this plant

9. Spelled Maiñ'gans in an earlier publication (Densmore 1913, 63).

in one column of her chart: "(1) pulverized," and on the next line she writes, "(2) Decoction." In the next column, describing how these medicines are administered, she writes, across from the pulverized description, "Snuffed up nostrils," and across from the decoction description, "Internally" ([1928] 1974, 340–41). As in this example, some of these entries describe more than one preparation method for the same plant part, and it is often difficult to be certain whether or not one is matching the preparation method with the correct plant because these two columns appear on opposite pages, not properly aligned. In addition, the recipes presented in this chart appear to be purposely abbreviated so that they will fit. Densmore uses this same chart format in other sections, although these other charts are not spread out over two pages. In entries for other, nonmedicinal, uses of plants, Densmore often includes recipes written in a much easier to decipher paragraph format. For example, Densmore gives the following recipe for a red dye made from "*Juniperus virginiana* L. Red cedar": "The bark of this tree was used by Chippewa women in Ontario for coloring the strips of cedar used in their mats.[10] A decoction was made of the dark red inner bark and the strips were boiled in it" ([1928] 1974, 371). This recipe appears to be somewhat clearer than those found in the medicine section. From this dye recipe readers know what part of the tree to use and how to prepare it. The information in "Uses of Plants by the Chippewa Indians" is not limited to recipes. Densmore also provides bits of cultural information about these plants, including two anishinaabe stories: "Legend of Winabojo and the Birch Tree," and "Legend of Winabojo and the Cedar Tree" ([1928] 1974, 381–86).

In *Chippewa Customs,* Densmore also provides quite a bit of information about how the Anishinaabeg use plants and trees. This is a much longer text than "Uses of Plants by the Chippewa Indians," but Densmore tends not to repeat the information she presents in the earlier text. She does not cover medicinal recipes in *Chippewa Customs,* although she does mention various cultural practices associated with the maintenance of medicinal knowledge, including the sweatlodge and the Midewiwin. She also provides an overview

10. Note: Densmore is speaking of two different trees in this recipe. Although the common name for *Juniperus virginiana* L. is red cedar, this tree is different from *Thuja occidentalis* L., commonly called cedar, the bark of which is used to make cedar mats.

of anishinaabe culture, and descriptions of various objects and practices associated with that culture. Plants and trees are used to make many of these objects, so they are an integral part of this text. Instead of giving specific recipes for medicines, foods, and dyes, as she does in "Uses of Plants by the Chippewa," Densmore provides instructions of varying lengths for making material objects, such as canoes, torches, and mats.

At the same time that Densmore was working on her research, Dr. Albert B. Reagan was researching gikendaasowin. While working as an agent for the U.S. Indian Service, beginning in 1899, Reagan conducted ethnological and archeological research among the native peoples where he was stationed, and he published his findings in a variety of journals, including the *American Anthropologist* and the *Wisconsin Archeologist* ("Notes and News" 1937, 187). In one of his assignments, he was stationed on the anishinaabe reservation at Nett Lake in Minnesota, where he conducted research among the Bois Fort Chippewa (Reagan 1922, 332). While there, he researched with George Farmer, an Indian policeman on the reservation. Farmer had a notebook filled with medicinal recipes and songs, which Reagan "accidentally" discovered. Reagan describes this event by saying, "Once when at his place I accidentally discovered that he had a large note book. His little daughter gave it to me, and on opening it, I saw writing in it, but in a language I did not recognize. After a good deal of persuasion, I succeeded in getting him to translate the words." (1922, 332). Reagan says that Farmer allowed him to copy the Ojibwe lines in this notebook, and he published medicinal recipes from this notebook in 1921 and the songs in 1922. In the article containing the recipes, Reagan presents readers with a line of the original Ojibwe, as he copied it out of Farmer's notebook, followed by a line of direct English translation. After presenting an entire recipe this way, Reagan gives an "Explanation" section, in which he presents more information about the recipe. Here is an example:

> Ki-sha-o-ti-sot a-ko-bi-son: (a) Ok-i-ni-mi-na-gash (b) ka-wa-go-mish,
> (For) cut foot apply on rosebush bitter root
>
> (c) mi-gwa-mi-ge-shi-na-gwag. Mi-squi-wit badj mi-na-a
> elm for bleeding little drink

> Explanation
> For a cut foot apply a tea made by boiling together roots of the rose-bush, bitter root, and elm. A little of this tea is also taken internally in cases of bleeding. (1921, 249)

As seen here, the plants in Reagan's articles are identified by common English names and Ojibwe names. As will be explained in chapter 2, keeping medicinal notebooks such as Farmer's is a common practice among members of the Midewiwin. The individual who creates one of these notebooks intends it for his or her personal use. Therefore, the recipes in them do not need to include step-by-step instructions, and in Farmer's, the recipes are clearly abbreviated. Here, we know that this tea should be applied to a hurt foot, but we do not know how strong a tea, or how long, or how often it should be applied. When presenting Farmer's songs in the other article, Reagan presents the entire song as written in Farmer's notebook and then divides it into separate lines, giving a "free translation" and a more literal translation of each line. He also explains when some of these songs are sung. All these songs appear to be Midewiwin songs; some of them are even labeled as such. In neither article does Reagan offer information about the original arrangement of recipes and songs in Farmer's notebook.

Reagan also published gikendaasowin in the *Wisconsin Archeologist,* including "The Bois Fort Chippewa," "Picture Writings of the Chippewa Indians," and "Plants Used by the Bois Fort Chippewa (Ojibwa) Indians of Minnesota" (1924; 1927; 1928). These articles also come from Reagan's fieldwork on the Nett Lake Reservation. In "Plants Used by the Bois Fort Chippewa," Reagan describes the source of the information he presents in this article: "For those who are interested in a medicinal way, the following general medicinal receipts are given as the writer found them copied in a medicine man's note book, the scientific names being added by the writer; the rest is given in direct translation" (1928, 231). Reagan does not name this medicine man, but these recipes are different from the ones in his previous article, which he says he copied from George Farmer's notebook.

Between 1932 and 1940, shortly after Densmore's and Reagan's last publications on gikendaasowin, Sister Mary Inez Hilger began conducting anthropological research among the Anishinaabeg. Hilger sent *Chippewa Child Life* to the BAE for publication in 1940, but it was not published until 1951 because of budget cuts brought on by World War II (O'Brien 1992, xv). Unlike Hoffman and Densmore, Hilger was trained in ethnology, and she held a Ph.D. in sociology, anthropology, and psychology from Catholic University, which she received in 1939 (Spencer 1978, 650).

As seen in the title *Chippewa Child Life and Its Cultural Background,* one of Hilger's research interests was childhood in various cultures. Robert Spencer, who wrote an obituary for Hilger, explains her research, saying, "Sister Inez saw the child as part of a whole cultural and social system. Her writings attest to her wish to depict the place of the child and the range of institutions touching childhood in selected social systems" (1978, 650). Along with *Chippewa Child Life,* Hilger published texts dedicated to the child life of other tribes (1951; [1951] 1992; 1952; 1957). Hilger says that she began studying the child in Indian tribes after finding that there was no such study available for any tribe at the time she was conducting her research (1960, i).[11] She explains her purpose for concentrating on "Chippewa" child life: "The purpose of this study is to record the customs and beliefs of the primitive Chippewa Indians of the United States as evidenced in the development and training of the child" ([1951] 1992, xxvii).

In *Field Guide to the Ethnological Study of Child Life,* Hilger describes the method she used to research the childhood of various tribes between 1932 and 1952 (1960, i, xi). This field guide provides a list of topics and associated questions for fieldworkers to ask native consultants about children and childhood. Hilger covers many of these topics in *Chippewa Child Life.* For example, under the topic of "Diapers" in her field guide Hilger writes,

> What substances are used as diapers (e.g., manure, fur, moss, inner barks, woven materials, or what)? Who prepares these (e.g., expectant mother or her mother, other relative, midwife, any woman)? When are they prepared

11. Hilger does acknowledge that Margaret Mead wrote texts with a similar subject, *Coming of Age in Samoa* and *Growing up in New Guinea,* but she says, "neither of these was like the study that I was undertaking" (1960, i).

> (e.g., just before birth, at any time, stored for future use and re-used)? How are they prepared (e.g., washed, sunned, smoked)? Are they ever obtained as gifts or purchases? (1960, 9)

In *Chippewa Child Life,* she gives details on some of these topics as they relate to the Anishinaabeg:

> Although rabbitskins were used as diapers, being washed and reused, or more often thrown away, swamp moss (āsȧ'kȧmik) alone or mixed with well-dried down of cattail, was most generally used. Moss was gathered from swamps and marshes in summer, hung on bushes until dry, and then shaken and pulled apart so as to rid it of all insects and dry weeds. Mothers kept supplies of it on hand, storing it in mākōk'. . . . "When moss was used for diapers the baby seldom became chafed, and when it was unwrapped you could smell only sweet moss." ([1951] 1992, 25)

As seen here, Hilger's field research methods resulted in very detailed information. Although focusing her research on childhood, Hilger does present information on a variety of topics, including, as seen in this example, various ways the Anishinaabeg work with plants and trees.

In 1940, the same year that Hilger submitted *Chippewa Child Life* for publication, Gerald C. Stowe published "Plants Used by the Chippewa" in the *Wisconsin Archaeologist.* No place in his five-page article does Stowe give a source of information, but it is clear that he copied this text entirely from Densmore's "Uses of Plants by the Chippewa Indians." Both Stowe and Densmore credit W. W. Stockberger, of the U.S. Department of Agriculture, with identifying sixty-nine plants used by the "Chippewa" that are also used medicinally by "white people" (Densmore [1928] 1974, 299; Stowe 1940, 8). Stowe claims he is citing information that Stockberger gained by working directly with the Chippewa, but he cites no text by Stockberger, and multiple bibliographic searches have concluded that Stockberger never published any separate texts with this information. Densmore, however, thanks Stockberger in her foreword for preparing this report for her to include in her publication and clearly states that Stockberger is describing the use of these plants by non-natives ([1928] 1974, 281, 299–303).

Throughout his article, Stowe gives the same uses for plants that Densmore gives, copying her sentences almost verbatim. For example, in his "Seasoning" section, a subsection under "Plants as Food," Stowe writes, "The red berries of the bearberry were cooked with meat as a seasoning" (1940, 11). In her section of the same title, Densmore writes under the entry "*Arctostaphylos uva-ursi* (L.) Spreng. Bearberry," "The red berries of this plant were cooked with meat as a seasoning for the broth. The leaves were smoked" ([1928] 1974, 318). Besides the obvious similarities in wording, this section is an example of plagiarism because here Stowe copies a mistake made in Densmore's research. I work with bearberry quite a bit, and I have eaten this plant's berries. They have very little taste, and I can only compare the texture of the berry itself to eating a synthetic sponge. This berry would provide no "seasoning" for any food; I guarantee it. Keewaydinoquay told Mary Geniusz that she had heard researchers say that the Anishinaabeg added bearberries to their soups and stews, but she insisted that bearberries are used as a means of extending an available source of food, not as a means of seasoning it. The bearberries, and not the food to which they are added, are being seasoned when used this way (M. Geniusz, pers. comm.). Citing Densmore's passage, Keewaydinoquay responds to the statement that bearberry is used as a seasoning by saying, "Chances are that the purposes were the reverse, i.e., the bearberry nutlets were added because of their nutritional value and the broth was intended to make the bearberries palatable" (1977, 8–9). Keewaydinoquay writes that Grandfather LoonHeart, an elder in her village, said that even the bears do not enjoy eating bearberries, and they only do so when they are starving and there is nothing else to eat. He admonished Keewaydinoquay and the other children listening to this teaching that they should remember how bear uses the bearberry, in case they too ever need to eat this food in times of starvation (Keewaydinoquay 1977, 8–9).[12] It seems that Densmore did not understand why the bearberries

12. Keewaydinoquay also says that her grandmother told her that when there was no other food available, the Anishinaabeg would fry bearberries in fat, mash them, and mix them with a few cut-up dried apples, and then flavor them with wild garlic, wild onion, yarrow, and wild basil. Keewaydinoquay adds to this recipe, "It required a family with strong teeth to masticate this cuisine" (1977, 10).

were added to the broth, and when he copied her research, Stowe copied this misinformation.

The final anthropological research cited in this text is Karen Daniels Petersen's "Chippewa Mat-weaving Techniques," published in a 1963 BAE bulletin. After this publication Petersen, a graduate of the University of Minnesota, published several texts on American Indian ledger art, and in her 1964 publication *A Cheyenne Sketchbook*, Petersen is listed as a staff member of the Science Museum in St. Paul, Minnesota (Cohoe 1964). The information in "Chippewa Mat-weaving Techniques" was the result of four research trips that Petersen made to anishinaabe reservations in Minnesota during 1957, 1961, and 1962. Petersen explains that she conducted this research because learning mat-weaving techniques from weavers was much more useful than studying woven samples found in museums. Obviously seeing her research as a means of preserving valuable information about mat weaving, Petersen writes, "in the foreseeable future such observation will no longer be possible among the Chippewa. The more difficult arts are dying with the older generation" (1963, 217). She adds to this statement that, even with the many acquaintances she made in anishinaabe communities over nine years, she had difficulty finding Anishinaabeg who still knew how to weave mats. Of those she did find she says, "All of these women were past middle age, and in no case did the younger woman who assisted the weaver know the technique before seeing it done for the purpose of research" (1963, 217).

In her article, Petersen gives detailed accounts of how the Anishinaabeg weave mats out of several different materials, including cedar bark, cattails, reeds,[13] and sweetgrass. She describes how the materials used to make the mats are gathered, prepared, and woven and provides diagrams for some of the weaving techniques she describes. She gives a brief biography for each "informant" she observed making mats. In this example, Petersen describes how to make a sweetgrass mat using a "coiling" technique:

> To begin a mat, the root ends of about 10 shoots, used without separating them into their approximately 16 separate leaves, are wound into as small a

13. Petersen identifies these "reeds" as *Phragmites communis* Trin. Var. *berlandieri* (Fourn.) Fern.

> coil as possible. A sewing needle is threaded with a single or double strand of lightweight cotton crocheting thread knotted at the end. Jones (1936, p. 27) specifies No. 10 thread to which beeswax is applied to make the sewing easier on the worker and on the strands of grass. Sewing is begun in the shape of the spokes of a wheel, each stitch passing around the outside of the coil and halfway through the opposite coil. This step as described by Jones (1936, p. 28) differs in beginning with knotting the end of the bundle and sewing the first coil to it. (1963, 274)

As seen in this passage, Petersen compares the techniques she observed while working with the Anishinaabeg to those observed by previous researchers. Here one also sees her attention to detail; rather than simply telling readers that the sweetgrass is formed into a coil and stitched into place, she describes the entire process of coiling. Someone who had strands of sweetgrass, but who did not know how to begin making one of these mats, could easily follow her instructions.

Spanning approximately seventy-two years, these texts show the evolution of colonized botanical gikendaasowin within the discipline of anthropology. They also exemplify varying degrees to which this information has been colonized and to which this information could be made useful to programs revitalizing izhitwaawin. Unlike the wordlists, non-natives, outsiders to anishinaabe culture, wrote all of these anthropological sources. While this alone does not make them colonized, the presentation of this information, and, as will be discussed in chapter 3, the fact that some of these sources contain degrading statements about the Anishinaabe and their culture does. These researchers have taken gikendaasowin and interpreted and presented it according to the expectations of anthropology during the times in which they were publishing. This is why many of these sources do not attribute pieces of information to specific consultants. In the late nineteenth and early twentieth centuries this attribution was not expected of an anthropological publication. One can see that by the time Petersen was writing, this practice was starting to change. The authors of these texts were describing the Anishinaabe and their knowledge to other non-natives. They were also actively trying to preserve pieces of Anishinaabe culture by writing about them.

## UNPUBLISHED ANTHROPOLOGICAL SOURCES

There is also unpublished anthropological research of gikendaasowin preserved in various archives. The earliest of these cited in this book comes from the papers of the WPA [Work Projects Administration] Indian Research Project, directed by Sister Macaria Murphy and conducted from 1936 through 1940 on the Bad River Chippewa Reservation in Wisconsin. Sister Mary Crandall, who wrote an introduction to the archived version of this research, says that this research was funded with a WPA grant, which Murphy received in 1935. Murphy was a college graduate who worked at Odanah, on the Bad River Reserve, from 1895 until 1946. Anishinaabe people conducted the research for this project, and Murphy assisted and directed them in their work (WPA 1936–40). The material gathered during this project was never published, but it was typed and compiled into a series of notes containing information about a wide range of topics within izhitwaawin. Some of these notes cite a person, or persons, who gave the information as well as a person who conducted the interview, while some of them just cite the person from whom the information came, perhaps because these individuals wrote their own notes. According to the description Crandall gives of this project, the people from whom this information came and the people conducting these interviews were all Anishinaabeg living on the Bad River Reservation. In this note, labeled "Pitch," Peter Marksman describes how the Anishinaabeg gather balsam pitch:

> In procuring this pitch the bark was chipped off with it and later the erudiate was removed. This mass was then boiled in water, during which process the pure pitch rose to the surface and was skimmed off. The pitch was again boiled until the substance assumed the consistency of resin when cool. In this condition it was ready to be stored for future use. When needed the application of heat rendered it pliable at once. The addition of fat or grease in varying quantities caused the pitch to remain soft and pliable even in very low temperatures. (WPA 1936–40, envelope 8, 45)

The paragraph quoted here is only part of a page of notes on balsam pitch. Marksman goes on to say that there are many uses for balsam pitch. It is used to heal cuts and scratches, as well as ringworms and other diseases of the skin. It was also used to make canoes watertight and to make torches (WPA 1936–40, envelope 8, 45). The notes found in the papers of the WPA Indian Research Project are of varying lengths, and many of them, such as this one, provide usable information on how the Anishinaabeg work with plants and trees.

The next group of unpublished anthropological materials on botanical gikendaasowin cited in this book comes from Erminie Wheeler-Voegelin's research conducted during the 1940s. Wheeler-Voegelin, who in 1939 became the first woman to receive a Ph.D. in anthropology from Yale University, was instrumental in the founding of the American Society for Ethnohistory (Tanner 1991, 62, 69). For two decades, beginning in the 1930s, she conducted research on various American Indian tribes with her husband, linguist Charles (Carl) F. Voegelin (Tanner 1991, 51–65). For one of their joint projects they conducted research in Ojibwe and Ottawa communities in and around Birch Island, Ontario, and Manistique, Michigan. During the summer of 1940, the two conducted a house-to-house survey of the community living at Birch Island, and during the summer of 1941 they returned to conduct linguistic and ethnographic fieldwork in this and surrounding communities under a Social Science Research Council grant. They intended to complete this research in the summer of 1942, but complications brought on by the United States' entry into World War II prevented them from doing so and ended this research, the results of which were never published (Tanner 1991, 63).

Among the professional books and papers that Erminie Wheeler-Voegelin gave to the Newberry Library in 1985 are her field notes from this research conducted with the Ojibwe and Ottawa in the Upper Great Lakes. As they exist now in the Newberry's Ayer Manuscript Collection, Wheeler-Voegelin's field notes from the summer of 1941 consist of four spiral notebooks filled with notes she took while talking to Anishinaabe consultants (EMV, folders 270, 271, 267, 268, 269). Information about these consultants is noted in the "Ethno-Linguistic Survey Summer—1941," by Erminie Wheeler-Voegelin and Carl F. Voegelin (EWV, folder 274). According to the names and locations written on the covers of these notebooks, Wheeler-Voegelin worked with three Anishinaabe consultants on this research.

Angeline Williams's name appears on the majority of Wheeler-Voegelin's field notes, filling two and one-half of the notebooks. Williams is "Marten" Clan and was born in Manistique, Michigan. She was sixty-four years old and living at Sugar Island, Michigan, when the Voegelins were conducting their research (EWV, folder 274). The name "Katherine Osogwin" appears on one of Wheeler-Voegelin's notebooks. Osogwin is "Manomeg" Clan and was born in Hessel, Michigan. She was approximately forty-nine years old and living in Hessel when the Voegelins were conducting their research (EWV, folder 274). Alec Boisoneau's name is written on one of Wheeler-Voegelin's notebooks, and information from Boisoneau fills half of this notebook. On the "Ethno-Linguistic Survey," only a Theophile Boisoneau is listed as a native consultant, and he is "Unicorn" Clan and was born on the Garden River Reserve in Ontario. He was seventy-three years old and living at Garden River during the summer of 1941 (EWV, folder 274).

Wheeler-Voegelin's notebooks are filled with botanical gikendaasowin. These notebooks contain information necessary for day-to-day living, such as this recipe for soup, given by Katherine Osogwin: "Whitefish—put on in cold water, + when it comes to boil, fish jiggled so later it will float on top of broth; when it comes to boil, it's done. Have cooked cornmeal in another pot; put fish on a dish; put cornmeal into fish broth + let it come to boil; then serve this soup first, + eat the fish next" (EWV, notebook 20). These notebooks also contain medicinal information, such as these medicinal recipes given by Angeline Williams: "Roots of Milkweed—when dizzy—used for heart trouble. Dried out of doors. Rubbed fine to powder. 1 tablespoonful to 1 qt. water, mixed; 1 whiskey glassful drunk. Take qt. in all. Pods put on hot ashes—breathe in smoke of them for heart trouble. Big medicine. Young shoots of [milkweed] not used" (EWV, notebook 19a). As seen in these examples, the gikendaasowin found in these notebooks spans a breadth of topics. Wheeler-Voegelin seems to have asked these Anishinaabeg many questions and taken copious notes on their answers. These notebooks are all handwritten, but Wheeler-Voegelin's handwriting is quite clear, and, as seen in these examples, her abbreviations are generally easy to decipher.

It is difficult to talk about specific "colonized" aspects of either the WPA or Wheeler-Voegelin notes because they are unpublished. No presentation of these materials has been made. Accessibility of these materials is limited to those with the resources to travel to locations where they are housed; however, it is commendable that they exist at all, given that they are just notes. They are mentioned here because they could be valuable resources for revitalization.

## ETHNOBOTANICAL SOURCES

Approximately four decades after the publication of the first anthropological source on gikendaasowin, ethnobotanists began publishing research on this topic. Although no single definition for the term exists, *ethnobotany* is the study of the place of plants in a people's culture (V. Jones 1941, 119–220; Schultes and Reis 1995, 11–12; Herron 2002, 8). Some ethnobotanists argue that the roots of their discipline can be traced back to Ancient Egyptian, Chinese, and Greek societies (Schultes and Reis 1995, 11, 20), but John Harshberger, a botanist from the University of Pennsylvania, first coined the term "ethno-botany" in a speech given in Pennsylvania before the University Archaeological Association on December 4, 1895 (Harshberger 1896, 146; Harshberger 1928, 4–5; V. Jones 1941, 219; Davis 1995, 41). In this speech, later published in the *Botanical Gazette,* Harshberger speaks briefly about the importance of "ethno-botany" as a means of clarifying whether a tribe practiced agriculture or primarily hunted, determining where plants once grew, showing ancient trade routes, and suggesting new uses for plants (1896, 146–53).

Huron H. Smith's *Ethnobotany of the Ojibwe Indians,* published in the Milwaukee Public Museum's 1932 bulletin, appears to be the first published ethnobotanical research on gikendaasowin. A scientist who did graduate work at Cornell University Smith was the curator of botany at the Milwaukee Public Museum from 1917 until he was killed in a car accident in 1933 (Barrett 1933, prelim.). While working at the Milwaukee Public Museum, Huron Smith began writing an ethnobotanical series, originally meant to include six volumes, on Wisconsin tribes. To write this series, Smith drew on his fieldwork, the objective of which was "to discover their present uses of native or introduced plants and, insofar as is possible, the history of these plant uses by their ancestors" (1932, 333). He published his research on the following four tribes in volumes of the Milwaukee Public Museum's bulletin:[14] the Menomini, the Meskwaki,[15] the Ojibwe [Anishinaabe], and the Potawatomi, which was published posthumously. Smith also conducted ethnobotanical fieldwork among the Winnebago and the Oneida (Barrett 1933, prelim iv (unnumbered); H. Smith 1923; 1928; 1932; 1933).

The information in the volumes of Huron Smith's ethnobotanical series is presented in a consistent format, each one divided into categories of use. The volumes contain sections on medicines, foods, fibers, and dyes. There are sections on miscellaneous use and medicine materials that are not plants. Smith also provides photographs of some of the plants of which he writes and of certain materials made from plants and trees. He acknowledges that his volumes, although each a book in itself, are only a sample of the knowledge maintained by these tribes. In the conclusions to his Potawatomi and Ojibwe volumes, he states that he has only been able to spend a few months with these tribes and that this short time has not been long enough to learn much of their ethnobotanical knowledge (1932, 433; 1933, 125). The following example, taken from *Ethnobotany of the Ojibwe Indians,* is typical of Smith's entries:

> Catnip (*Nepeta cataria* L.) "tci' name' wûck" [big sturgeon plant]. The Flambeau Ojibwe brew a tea of catnip leaves for a blood purifier. The mint water obtained by steeping the herb in lukewarm water is used to bathe a patient, to raise the body temperature. The plant is employed by whites as an emmenagogue and antispasmodic. It has been used as a carminative to allay flatulent colic in infants, and is supposed to be useful in allaying hysteria. (1932, 372)

As in many of his other entries, here Smith identifies the plants in his texts by scientific, common, and native names. He gives several sentences on how the plant is used by the native people of whom he is writing. He also notes whether non-natives use this plant.

14. All of these tribal names are spelled according to how Smith spells them in his volumes.

15. H. Smith notes that the Meskwaki do not live in Wisconsin at the time he is researching, but he says that he includes them because they once did (1928, 180).

Four of the ethnobotanists who worked with the Anishinaabeg were associated with the University of Michigan's Ethnobotanical Laboratory. The earliest anishinaabe ethnobotany produced by the University of Michigan is Melvin R. Gilmore's "Some Chippewa Uses of Plants," published in 1933. Gilmore, who held a doctorate in botany, was the first curator of ethnology for the Museum of Anthropology at the University of Michigan. He was also one of the founders and the first director of the Ethnobotanical Laboratory (Cowan 1988, 456; DiBella 2004). Gilmore says that he and his cofounder first saw the need for the Ethnobotanical Laboratory during the summer of 1930, when several archaeologists asked Gilmore to identify botanical materials from their fieldwork. They decided that a place was needed where archaeologists could send botanical samples for proper identification because, as Gilmore writes, "the development of a people's phytotechnic is fundamental to the development of all their arts of life" (1932a, 12). He argues that a people cannot be understood if one does not know both the geographical and the botanical features of that people's environment (1932a, 17). Gilmore clearly saw ethnobotany as an important component in any ethnological study, and he wrote "Importance of Ethnobotanical Investigation" as a helpful guide for anyone undertaking ethnological fieldwork (1932b, 327).

Gilmore's "Some Chippewa Uses of Plants" consists of a list of 115 plant species, which he divides into 49 plant families. Under the identification information for each plant are short descriptions of how the Anishinaabeg use, think about, or interact with that plant. These descriptions are short, ranging from one sentence to three paragraphs in length, and to his credit he does note that "Chippewa Uses of Plants" is only the beginnings of an ethnobotanical study (1933, 120). By the mid-1930s, Gilmore was in very poor health (Cowan 1988, 456), which may explain why he never published a lengthier report on anishinaabe ethnobotany. Here is a sample of one of Gilmore's longer entries:

> *Thuja occidentalis* L. Arbor Vitae. Kizek, Kiskens, Kisgens, Songup
> Twigs of arbor vitae were burned as a disinfectant to fumigate a house in which one was sick of a contagious disease, such as smallpox. Many years ago when smallpox first came to the Chippewas they moved into arbor vitae swamps and camped there during the period of the plague. Twigs were used in the vapor bath, and were burned for incense in religious ceremonies

> and as a common deodorant. The leaves were infused in hot water to make a beverage like tea; they were also used for perfume for the clothing and combined with ground hemlock for medicinal purposes. Ribs of canoes, as well as toboggans and handles of sturgeon spears, were made of arbor vitae wood. (1933, 123)

As seen in this entry, Gilmore provides several ways to identify each plant, including scientific, common, and Ojibwe names. From his other article it is clear that he views proper plant identification as an essential component of the ethnobotanical investigation (1932b, 325–27). In this entry, Gilmore includes physical uses for this tree, such as making ribs of a canoe or making tea, but he also writes about the spiritual connection between this tree and the Anishinaabeg. Gilmore is critical of ethnobotanists who were only seeking "economic" uses for plants, such as making food, textiles, and tools, and he argues that ethnobotanical investigation should be a mix of this economic botany along with the "whole range of knowledge of plants and plant life" (1932b, 322). Gilmore was also interested in Ojibwe plant names, and lists a few of them in his article. He encourages others working with native plant knowledge to do the same as these names may "disclose significant facts not otherwise discoverable" (1932b, 327).

Volney Hurt Jones was also an ethnobotanist at the University of Michigan, and he became the director of the Ethnobotanical Laboratory upon Gilmore's death in 1940 (Cowan 1988, 456). Volney Jones's career at the University of Michigan began in 1933 and continued until shortly before his death in 1982, although he retired in 1969 (Cowan 1988, 456–59). Jones held an M.A. from the University of New Mexico with a biology major and an English minor (Griffin 1978, 3–4). He conducted fieldwork among Indians in Michigan and Ontario during 1933 and 1934, and this fieldwork resulted in several publications on the ethnobotany of the Anishinaabeg, including one describing how the Anishinaabeg make brooms, another noting their uses for sweetgrass, and two describing different kinds of woven mats (Griffin 1978, 4, 11; V. Jones 1935; 1936; 1946).[16] Jones coauthored an

16. Jones's final article containing anishinaabe ethnobotanical information, "The Bark of the Bittersweet Vine as an Emergency Food among the Indians of the Western Great Lakes

article with Vernon Kinietz on Anishinaabe consultants making rush mats (Kinietz and Jones 1941).[17]

In "The Nature and Status of Ethnobotany," Volney Jones outlines advice for how ethnobotanical investigations should be conducted, which he follows in his articles on gikendaasowin. Criticizing other ethnobotanists for simply describing how native peoples used plants, he argues that the purpose of ethnobotany is to look at "the entire range of relations between primitive man and plants" (1941, 219). Jones's articles on anishinaabe ethnobotany differ from those texts by Huron Smith and Melvin Gilmore in that rather than listing many plants and briefly stating ways in which the Anishinaabeg use those plants, Jones focuses each of his articles on one plant or one cultural practice. In "Some Chippewa and Ottawa Uses of Sweet Grass," Jones provides a detailed account of how the Anishinaabeg tend, gather, prepare, and use sweetgrass for prayer and to make baskets and ornaments (1936). He also provides photographs of the finished products. For example, after describing all the materials used to make a sweetgrass basket, even describing how the thread is drawn through beeswax before being used, Jones writes,

> The workman is seated and has a table or other support for materials and equipment. The needle is threaded, and one end of the thread is knotted, so that it may be used for single-strand sewing. A small bundle of the grass about one fourth of an inch in diameter is taken; the ends are evened and the coarse bases cut off with the scissors. If the coil is to be entirely of grass, the base end of the bundle is knotted by looping the end over the bundle and pulling it through. The first coil is started by wrapping the long end of the bundle around this knot. The sewing is first through the knot, looping

---

Region," is written entirely from secondary sources and from Jones's own experiment of cooking and eating this bark (1965).

17. This article is typical of Jones's other articles on gikendaasowin. Although the authors state that neither of them watched the entire process of making a rush mat, together their notes in this article provide step-by-step instruction on processes related to making rush mats. Kinietz published other research among various Indian tribes, including *Chippewa Village: The Story of Katikitegon* and *The Indians of the Western Great Lakes, 1615–1760,* but neither of these contains very much information about botanical gikendaasowin (1947; 1965).

> the thread around the long end of the grass, and back through the knot. As each stitch is made the grass is bent firmly against the knot and the thread pulled tight. After the grass has completely encircled the knot, the sewing is into the previous coil instead of into the knot. (V. Jones 1936, 28)

Jones continues with this detailed description, even explaining which fingers and which hands hold which parts as the process continues. From this description, it is clear that he actually watched this process and took careful notes. He does not simply tell readers to coil strands of sweetgrass; he explains, step by step, how this coil is made. Jones argues that the ethnobotanist's job is to negotiate between the fields of plant science, plant ecology, and anthropology, presenting research that will be of interest and use to researchers in all of these fields. He adds that ethnobotanists need to be "familiar with the techniques, methods and approach of both anthropology and the plant sciences," and he encourages continued cooperation between these fields (1941, 220).

Keewaydinoquay, who occasionally published under her English name, Margaret Peschel, was also an ethnobotanist at the University of Michigan.[18] According to her academic advisor, Richard I. Ford, she completed all of the coursework for a doctorate of philosophy in biology from that university (pers. comm.).[19] As stated previously, Keewaydinoquay also held a master's degree in biology from Central Michigan University. It is hard to place Keewaydinoquay's work strictly within the category of anishinaabe ethnobotany as she was both an academic and a *mashkikiiwikwe,* and as such her work with gikendaasowin really falls into two categories: academic preservation and community maintenance of this knowledge. During her lifetime, Keewaydinoquay published several short books on the ethnobotany of the Anishinaabeg, including *MukwaMiskomin or Kinnickinnick: "Gift of Bear": An Origin Tale Never Before Recorded How to Use Bearberry for Teas, Emergency Food, Treating Diabetes and Internal Infections, Making Non-narcotic*

18. For more information on Keewaydinoquay's life, see Peschel 2006 and W. Geniusz 2005.

19. Ford says that Keewaydinoquay did not complete this degree because she was unable to pass a proficiency test in German (pers. comm.).

*Smoking Mixture* (1977). She also made audiotapes of some of her class lectures at the University of Wisconsin-Milwaukee, where she taught courses on ethnobotany and the philosophy of Great Lakes Indians, and made videotapes of her lectures at ethnobotany and philosophy workshops sponsored by the Miniss Kitigan Drum. In each of her texts and lectures, Keewaydinoquay focuses on one plant or one plant family. Ford describes how her ethnobotanical studies differ from those of most ethnobotanists:

> She is deeply interested in how people relate to plants and respect them, not just how they use them as is traditional in most ethnobotanical studies. As such, her ethnobotany is different from most and that is what gives it special significance, unusual quality, and unparalleled importance. Unfortunately, many ethnobotanists concentrate on examining the utilitarian value of plants and forget, or do not understand, their meaning in the lives of the people. Keewaydinoquay is an Ahnishinaabeg (Chippewa) elder who wants us to learn ethnobotany, not from a western, scientific perspective, but by appreciating the botany of a people—a deeper and more significant meaning of ethnobotany—truly the botany of the people. (1998, ix)

Keewaydinoquay's ethnobotanical teachings focus not only on how the Anishinaabeg use a particular plant but also on many teachings about that plant from the perspective of izhitwaawin. She begins *Mukwamiskomin or Kinnickinnick: "Gift of Bear"* with the origin story of *mukwamiskomin,* also called *Arctostaphylos uva-ursi* (L.) Spreng, or Bearberry, which MidéOgema, her grandfather and a leader of the Midewiwin, told to her as a child (1977, 1–3). In this same text, Keewaydinoquay presents four teachings from Nodjimahkwe, her Anishinaabe teacher. In one of these teachings, Keewaydinoquay recounts Nodjimahkwe's words:

> Now to speak of one which is very important for our people, the trouble of sugar in their water or slow death when ants come to their urine. Make of the Blueberry leaves (Vaccinium sp.) and fruits, at the time they are ripe, the tincture, as I have shown you. They will preserve themselves because of the ishkotewabo. Make also a tincture of Mukwa-Miskomin leaves and berries; this one is a little easier because you can do it any time from August through November. Have the ones so affected (by diabetes mellitus) prepare

> a cup of spring water which can be cold or hot—not boiling. To this they must add 20–40 drops of the blueberry tincture and 10–20 drops of the Bearberry tincture, depending on the need of the case. No more at one time, ever! This should be taken 2 or 4 times a day. (1978, 23)

In this passage, Keewaydinoquay provides a recipe with enough details to enable readers to make this medicine. This recipe provides dosage information as well as a warning about not taking too much at one time. Keewaydinoquay also provides notes for her readers to clarify certain parts of the recipe, including an explanation that "slow death when ants come to their urine" is diabetes mellitus. As Ford says, Keewaydinoquay's ethnobotanical teachings do not stop at usage information. She also presents these teachings within the context of izhitwaawin. After this passage, for example, Keewaydinoquay provides cultural information about this medicine, including how she was to learn it within the context of izhitwaawin by reciting this recipe to Nodjimahkwe every day "until the next full moon has come and gone" (1977, 23).

In her writings and recorded lectures, Keewaydinoquay also uses her non-native training as an ethnobotanist to embellish and strengthen the cultural teachings that she provides. In *Min: Anishinaabe Ogimaawi-minan: Blueberry: First Fruit of the People,* Keewaydinoquay gives a dibaajimowin about blueberries that she writes in "hieroglyphs" and Ojibwe using Romanized letters (see illustration 1 in chapter 2), providing English translations for both. This dibaajimowin says that a family who dries plenty of blueberries during the Blueberry Moon will walk in strength in the spring (1978, 33). With this dibaajimowin about blueberries, Keewaydinoquay presents an analysis of this teaching, partially based on scientific information. She states, "BlueBerries contain a high proportion of macronutrients and micronutrients which are necessary to maintain good health" (1978, 33).

As seen from these sample entries, most of these ethnobotanical texts contain detailed descriptions about how the Anishinaabeg work with plants and trees. They could be invaluable resources for a revitalization program. Some of them also include Ojibwe names, which, as discussed in chapter 3, could be worked with to produce even more resources. With the possible exception of Keewaydinoquay's text, these sources are all colonized to varying degrees. As discussed in chapter 3, some of them contain degrading comments about the Anishinaabeg and their knowledge. All of them present information, in some cases quite detailed, according to the philosophies and knowledge-keeping systems of the colonizers. In most cases, they discuss the physical properties of these plants, while mentioning briefly or ignoring entirely the spiritual ones. One might categorize Keewaydinoquay's text as colonized as well because although writing from the perspective of izhitwaawin, she includes scientific information, making it appear as though she writes from the perspective of the colonizer. Some will argue that this makes her work colonized; others will disagree. Once again, debating what should or should not be labeled as a "colonized" text takes our focus away from the real issue—how can we bring this information back into anishinaabe communities? Labeling aside, what it comes down to is the desires of the specific program trying to use this knowledge. Many programs wish to teach this information orally to participants in the form of outdoor lectures, so this information will end up being reworked no matter who wrote it. In such cases, individuals may choose to ignore the scientific data Keewaydinoquay presents and simply not include it in their presentations. Chapter 3 contains a longer discussion on the importance of not ignoring gikendaasowin simply because it has been influenced by non-native knowledge-keeping systems. Readers should take caution, however, and realize that the goal of this discussion is to show what is out there, not to label sources as "colonized" and to ignore them completely. Such finger-pointing takes our focus away from the real goal here: to decolonize our gikendaasowin so that it may help our people regain their strength.

## RECENT SOURCES

Over the last two decades authors from various backgrounds and disciplines have published texts on botanical gikendaasowin. One example is *Wild Rice and the Ojibway People*, by Thomas Vennum, Jr., who acknowledges that when researching this text he went outside of the topics generally researched by those in his field of ethnomusicology (1988, ix). In discussing previous publications on wild rice, Vennum notes that the two existing publications by Albert E. Jenks and Eva Lips are "generally inaccessible and both need updating" (1988, viii). In comparison, Vennum's publication is definitely accessible, being still in print after almost twenty years, and it is often available for sale at pow wows in the Midwest. He provides a detailed account of contemporary as well as historical wild rice harvesting and its context within izhitwaawin. Looking at this topic from many angles, Vennum discusses everything from the traditional harvest of wild rice and its importance to izhitwaawin to a more scientific look at the plant itself and a nutritional breakdown of vitamins and minerals found in wild rice. His research draws on numerous published and archival resources and interviews with contemporary Anishinaabe wild rice harvesters. Vennum's text might serve as a model for a decolonized text. Yes, a non-native wrote it. Yes, like Keewaydinoquay's texts, it does include some scientific information. It also includes important gikendaasowin, which is presented in a respectful way. Vennum worked with Anishinaabe elders on this research, and he gives them proper credit for sharing their teachings with him.

Soon after the publication of Vennum's text, Ron Geyshick, a medicine man from the Lac La Croix Reserve in Ontario, wrote *Stories by an Ojibway Healer: Te Bwe Win: Truth* with Canadian filmmaker Judith Doyle. In her introduction, Doyle explains that the stories that Geyshick tells in this book are his own stories, and that she has not changed these stories for publication. She says that when she met Geyshick while working on a documentary film at Lac La Croix three years before the publication of this text, he had already begun writing this book. She adds, "It was important for me not to change the form of his stories" (Geyshick 1989, 7). Geyshick writes about gikendaasowin and shares a few medicinal recipes with his readers. Geyshick clearly wants to share all of the information he presents in his book with future generations. He writes, of the medicines in his book, "These are important medicines to know, and I want to pass them on. Anyone who picks this medicine should leave some tobacco by the tree and tell the spirits I told them to come and ask for this assistance" (1989, 27). Geyshick also shares other information about izhitwaawin, including dibaajimowin and information about being a healer, in this text. Here is one of the recipes that Geyshick shares with his readers:

> If a woman falls during her pregnancy and starts feeling pains, there is nothing white doctors can do for that. But we have a medicine. Someone should

> run right into the bush and find a little balsam tree—how big depends on how many months along the woman is. At two months, you should pull out the smallest tree, roots and all. At four months, pick one maybe six inches high. And at six months it should be around a foot. The further due she is, the bigger a tree you pull out. Boil the whole thing, roots and all, and have her drink this. (1989, 26)

As seen in this example, Geyshick gives these recipes and provides details as to the dosage that should be used in different cases. He usually identifies the trees and plants by English common name, and in cases where he does not know an English name for a plant, he provides a description of the plant in question and its growing conditions. It is clear from his descriptions that Geyshick has made these medicines himself. For example, when giving the recipe just quoted he writes, "Now in the winter, it can be hard to see these tiny balsam trees under the snow, but you can use a bigger one instead. Break the very top off, about one year grown, and boil it. Often where there are bigger trees you'll find the little ones, if you dig under the snow" (1989, 26). Here Geyshick provides instructions for possible problems that one might encounter when making this medicine. One who had no experience making this medicine would most likely not be able to predict this problem and readily offer a solution to it.

Another recent text on anishinaabe plant knowledge is the Great Lakes Indian Fish and Wildlife Commission's [GLIFWC] *Plants Used by the Great Lakes Ojibwa*, by James E. Meeker, Joan E. Elias, and John A. Heim.[20] This text is still available through GLIFWC and is for sale at many pow wows. It is a reference work, which provides identification information for plants growing in the Great Lakes area, along with very brief descriptions of how Indians, not necessarily the Ojibwe, have used these plants. Although in the

20. An abridged version of *Plants Used by the Great Lakes Ojibwa* was also published in 1993 by GLIFWC, containing only the original introductory pages and the tables of plant names (Meeker, Elias, and Heim 1993b). All of this material appears exactly as it does in the original text, except for a change in the pagination. The tables found in this text list Ojibwe, common, and scientific names for all of the plants presented in the unabridged text, but they include neither the diacritics nor the standardized plant names found in the body of the unabridged text.

preface the authors say, "Each of the plants named in this book was used by the Anishinabe *[sic]*," it is questionable what tribe the authors reference in the descriptions of how some of these plants are used, and they do not cite specific sources in their entries. Their entry for *Sambucus racemosa* (red elderberry) exemplifies this: "The fruits are used in jellies, wines, and pies, but the vegetative parts are poisonous. In Native American medicinal practices a decoction of inner bark was used as an emetic or cathartic although it was considered dangerous. An infusion of the root was also used for unspecified medical purposes" (1993a, 305). From this entry it is unclear to what tribe Meeker, Elias, and Heim are referring. All the reader sees is the broad category of "Native American." It is also unclear who makes "jellies, wines, and pies" out of this plant. As this example demonstrates, one could not actually use any of the plants with just the brief descriptions provided in this text. To their credit, however, it should be noted that Meeker, Elias, and Heim are some of the few authors who include warnings about poisonous plants.

The authors of *Plants Used by the Great Lakes Ojibwa* also include Ojibwe names in their text, such as these listed for *Nymphaea odorata* (the white water lily): "**Akandamoo** (Baraga: akandamo, -g 'a kind of big root growing in the water'; Rhodes: kandmoo) [Smith: odîte'abûg wa' bîgwûn, odîte'abûg wabî'gwûn; Zichmanis and Hodgins: anung pikobeesae]" (Meeker, Elias, and Heim 1993a, 143). As seen here, some of these entries include plant names written in a standard orthography, presented in bold font. John Nichols, who collected the Ojibwe names presented in this source, explains that he collected these names from Maude Kegg and Edward Benton-Banai, two Ojibwe speakers. In some cases, Kegg or Benton-Banai supplied the name listed and in others, they verified a name found in a written document. Plants for which there is no standardized Ojibwe name are plants that neither Kegg nor Benton-Banai recognized. Nichols explains further that, as seen in this example, immediately after a standardized name are similar names for that plant written as they appear in other sources, followed by other Ojibwe names for that plant (pers. comm.). The introduction also explains that it was not possible to figure out all of the names in this text, and these names are left as the original researcher wrote them (1993a, 1). There are some entries, such as those for *Ranunculus sceleratus* (the cursed crowfoot) and

*Rumex altissimus* (the water dock) for which the authors provide no Ojibwe name at all (1993a, 180–81).

The last six years have seen the production of several new ethnobotanical texts on gikendaasowin. In a recent dissertation, Scott M. Herron evaluates retention of plant knowledge in Ojibwe, Ottawa, and Potowatomi communities in Wisconsin, Michigan, and Southern Ontario. He also identifies plants known to individuals in these communities and briefly describes their use (2002). Recent ethnobotanical investigations have also looked at categorization of plants and trees within the Ojibwe language. A master's thesis by Mary B. Kenny (2000) and an ethnobotanical article by Ian J. Davidson-Hunt, Phyllis Jack, Edward Mandamin, and Brennan Wapioke (2005) both compare Brent Berlin's theories of universal folk taxonomy to Ojibwe categories of plants. In their respective works, which concentrate on separate communities of speakers, these authors conclude that, although in some ways Ojibwe categories do fit into this system, in other ways they do not. Davidson-Hunt, Jack, Mandamin, and Wapioke further suggest that a "holistic" approach, rather than one that looks exclusively at plant names and classification, should be used when studying anishinaabe ethnobotany as plants are an integral part of the anishinaabe way of life (2005).

## CONCLUSION

When looking at gikendaasowin as it has been preserved within the academic record and as it is maintained by the Anishinaabeg, it is important to realize that these are two totally different information-keeping systems, which have preserved, in different ways, what was once the same knowledge. Various researchers elicited botanical information from the Anishinaabeg. This information was collected and written according to the guidelines dictated by these researchers' backgrounds, professions, and places of publication. We do have some texts containing botanical gikendaasowin written by Anishinaabe authors, but only a few exist and even the ones appearing to be less influenced by colonization are so short that they contain little information. We are now in a time of great cultural revitalization among the Anishinaabeg and other Indian peoples. Individuals, schools, and various organizations are attempting to use written sources documenting gikendaasowin in their revitalization programs. For many reasons, including the background of the researcher, the era in which he or she was writing, the format of the publication, and the amount of time the researcher spent conducting his or her research, the presentation of the information in many of these sources makes them inadequate tools for the revitalization of izhitwaawin. For the most part, even in the case of sources written by authors who took very detailed, careful notes about what they were researching and who published or otherwise preserved step-by-step instructions on how to work with certain plants and trees, the knowledge-keeping systems out of which these sources were written conflict with those of gikendaasowin.

# 2

# Botanical Anishinaabe-gikendaasowin Within Anishinaabe-izhitwaawin

## INTRODUCTION

The first step in Biskaabiiyang research is to return to our teachings. Only from that perspective can we begin to decolonize our knowledge and ourselves. When decolonizing written accounts of botanical gikendaasowin, therefore, we need to know to where we are returning; that is, we need to know how izhitwaawin maintains gikendaasowin. Part of this process requires the researcher's internal self-reflection, and part of this process requires new external research. By piecing together information from colonized texts and comparing it with information coming from anishinaabe sources, we can understand how knowledge originates and is maintained within izhitwaawin. Once we have this understanding, we will be in a better position to search through these texts of colonized botanical gikendaasowin and make decisions about which pieces are usable, which pieces need to be elaborated, and which ones need to be discarded.

As explained in the introduction, we as Anishinaabeg have our own reasons for conducting research on our culture. Rather than trying to explain ourselves to the rest of the world, we are trying to regain and revitalize teachings that were or are being lost from our families and communities. Biskaabiiyang research methodologies can help us achieve this goal by providing us with a common language through which researchers and community members can come together to discuss information and a means to decolonize ourselves so that we may accept and respect the information we receive through oral and written sources.

Biskaabiiyang research begins with the researcher, who must evaluate the effects that colonization has had on his or own life and mind and, having recognized those effects, must return to our original teachings so that he or she can conduct meaningful research for our people. A lot of concepts and teachings in this chapter, when viewed through the eyes of the colonizer, appear "fictional," "childish," "not scholarly enough," or just plain "ridiculous." For instance, the importance of someone conversing with a plant and making an offering before picking it, although something I have done all my life, is not generally included in a "scholarly" text. Some academic texts describe this practice as something that Anishinaabeg do, but it is not discussed as an essential part of the research process. As someone who was raised in both the dominant culture and the anishinaabe culture, I had to see past the colonized portion of my mind before I could write this chapter.

Biskaabiiyang research methodologies use informative sources differently than many other methodologies. This Biskaabiiyang research is an attempt to regain information so that it may be used to revitalize izhitwaawin; therefore, both oral and written sources are equally accepted. In some academic texts, written sources are used as a means of verifying an oral source, the accuracy of which is often questioned. One might argue that Biskaabiiyang research, at least the way it is used in this book, places a higher importance on oral teachings from our elders than on written sources. In this chapter and in chapter 4, written and oral sources are used together to help us learn all we can about these practices so that we may use them in our own lives. Using multiple sources is particularly helpful when talking about certain teachings and pieces of information that are no longer a part of our everyday lives because these are the kinds of things that are most in danger of being lost from our communities.

## PLANTS WITHIN IZHITWAAWIN

It should be noted that there are no words for "plant" or "botanical knowledge" in Anishinaabemowin, although there are names for different plants and various ways one can describe certain kinds of plants. There is a word for tree, *mitig*, but whether or not everything we think of as a "tree" in English falls into this category is a matter to be debated. Although one can describe wanting to know about how the Anishinaabeg use a mitig, the concept is alien to gikendaasowin because the use of trees and plants is not a category prescribed by gikendaasowin. Instead, there are things that one learns within the context of izhitwaawin, and these various things require learning about how to use, work with, and ask for the assistance of plants and trees. To make certain objects, such as shelters and canoes, or to prepare foods and medicines requires a certain amount of knowledge about working with plants and trees. There are also certain spiritual understandings about plants and trees that are necessary to participate in izhitwaawin. Therefore, although "plant knowledge" or "tree knowledge" are not terms that readily translate into Anishinaabemowin, having this knowledge is essential to many aspects of inaadiziwin and izhitwaawin. As will be explained in this chapter, botanical knowledge is an integral part of inaadiziwin and the decolonizing process.

## MASHKIKI WITHIN GIKENDAASOWIN

There are other categories of gikendaasowin into which botanical knowledge fits, such as *mashkiki* (medicine), but not everything within that category comes from botanical sources. When I first explained my research to Dora Dorothy Whipple, she responded by saying that if I was going to research plant medicines, I should also include spring water because within gikendaasowin it too is considered a mashkiki. During one of our recording sessions, she told me more about the importance and origin of *mookijiwaninibiish*,[1] spring water:

> Mii giiwenh eta gii-noondamaan iko mewinzha, chi-mewinzha . . . Anishinaabeg bwaa-gikendamowaad i'iw nibi. Gaazhi-bawaajiged aw inini ashi-nibo, aw inini aakozi. Gaazhi-bawaajiged iw. Minikwed i'iw. Giiwenh da-mino-ayaa. Mii gaazhi-maajaawaad iniw gii-ozhiwanikewaad gii-nandonendamowaad iw nibi . . . gaazhi-mikamowaad i'iw spring water Mii maa gii-kabeshiwaad. Miish iw gaa-minikwed ayiidog aw inini ashi-nibod aakozid. Mii gii-pawaajige giiwenh i'iw minikwed i'iw ji-mino-ayaad. Mii

1. This word comes from George McGeshick, Sr. (pers. comm., July 30–Aug. 1, 2005). Whipple could not remember the word for spring water.

> geget gaazhi-minikwed i'iw, mii maa gabeshiwaad Anishinaabeg. Ashi-mino-ayaad. Miish apane ezhi-noondamaan igo gaazhi-noondamangwaa igo Chi-anishinaabeg gegoo iw ogii-manidoowendaanaawaa i'iw nibi, gii-minikwewaad iw. Chi-waasa ko giiwenh gii-izhaawag iniw gii-nando-nendamowaad iw nibi. (pers. comm.)

Here Whipple says that a man who was very sick, almost dying, dreamed about spring water. When he drank it, after he awoke, he was well. She adds that the Anishinaabeg used to dig holes looking for spring water, and they would camp where they found it. The Chi-anishinaabeg, or Gichi-anishina-abeg (old-time Indians), saw spring water as a spiritual thing, she adds, and they would drink it. They would go far away to find it. Before telling this story, Whipple says that she herself uses spring water.

Mary Geniusz says that Keewaydinoquay had certain medicines that she would only make if she had water from a specific spring in Michigan. One of these medicines Keewaydinoquay called a "Spring Tonic." Geniusz says she does not know if this name comes from spring water, one of the key ingredients in this medicine, or from the fact that people often drink this in the springtime. Keewaydinoquay taught that after a winter of eating salty foods and not being very active, a person's blood would thicken, and this mashkiki would thin a person's blood and clean out his or her system, both very important things to do to prepare a person's body for the activity of spring and summer. Geniusz says this medicine is also given to people who have circulatory problems, diabetes, and certain heart problems (pers. comm.).

There are other items as well, which although considered mashkiki are also not derived from botanical materials. Huron Smith lists rattlesnake flesh, white clay, and bear fat as being items that are mixed with plants to make various mashkiki. He says that white clay alone, which he calls "*waba'bîgan,*" is also a mashkiki (1932, 352). Kathryn Osogwin says that when a little skunk oil is hung on the door, disease will not come to that area as long as a little of the odor remains (EWV, notebook 20). Whipple says that when she was a child sometimes a little bit of skunk oil was used to treat her siblings and her if they were sick (pers. comm.). Mary Geniusz says that Keewaydino-quay also spoke of how the Anishinaabeg use skunk oil to treat a patient with pneumonia or another form of heavy congestion. Only a very small amount of skunk oil, approximately one drop mixed with cooking oil, is used to treat heavy congestion. Keewaydinoquay told Geniusz that when she was a child people often used a rabbit skin that had been sprayed by a skunk, or one that was rubbed on something that had been sprayed, as an inhalant in place of straight skunk oil (pers. comm.).[2]

It is important to understand these categories so that one can begin to understand how botanical gikendaasowin fits into inaadiziwin and izhitwaawin. Botanical gikendaasowin touches many areas of izhitwaawin, and the concepts that apply to trees and plants, such as those that will be described in the next section, are only a small part of inaadiziwin.

## CATEGORIES OF ANIMATE AND INANIMATE WITHIN INAADIZIWIN

The protocols of izhitwaawin dictate that we must make an offering of *asemaa* (straight tobacco)[3] or a mixture of herbs, often written in English *kinnickinnick,* when asking the assistance of another being. For example, Whipple tells a dibaajimowin about the importance of making an offering of asemaa during a thunderstorm to ask the *animikiig* (thunderbirds) to pity the Anishinaabeg living in that place and not create weather conditions that could harm them (Whipple 2006a). When working with us on the CD-ROM *Asemaa: Tobacco,* Ken Johnson, Sr., recorded seventeen teachings describing different times when it is necessary to make an offering of asemaa to different beings. For example, one must make an offering to the beings in a body of water before crossing it and to an elder from whom one wishes to hear a story (Johnson 2006). These offerings are necessary because of the structure of inaadiziwin, which views the world differently than most non-native philosophies and ways of being.

2. Keewaydinoquay preferred other decongestants, such as mint, to the heavy odor of skunk oil. She lectured about skunk oil in her university classes when lecturing on skunk cabbage (*Symplocarpus foetidus* [L.] Salib. ex Nutt.) because both the cabbage and the oil of the animal are used the same way to treat heavy congestion (M. Geniusz, pers. comm.).

3. Keewaydinoquay identified asemaa as *Nicotina rustica,* a different species than the plant used in the production of commercial cigarettes (M. Geniusz 2005, 15).

Inaadiziwin divides the world into categories of animate beings and inanimate objects differently than other philosophies. Insisting that one cannot understand why the Anishinaabeg "have so much faith in the healing of the plants" without understanding anishinaabe "general philosophy," Keewaydinoquay says that things such as rocks, trees, and plants, which are considered to be inanimate by the dominant society, are considered animate within anishinaabe philosophy. They, along with all the other beings of Creation, have a special purpose, which they must follow to maintain the balance of this world (1991b). We can find linguistic evidence of this dibaajimowin within Anishinaabemowin, the nouns and verbs of which are divided into animate and inanimate categories. A verb describing the actions of an animate noun cannot be used to describe the actions of an inanimate noun. Therefore, the same verbs that are used to describe the actions of a human being are also used to describe the actions of other animate beings, such as rocks, trees, and certain religious items. For example, when speaking about the importance of asemaa and describing different times that the Anishinaabeg make offerings of asemaa, Ken Johnson, Sr., says, "Asemaan odaabaji'aan Anishinaabe gii-ando-madoodood. M'apii waabamigoowizid Anishinaabe owe biindaakoonaad owe gimishoomisinaanig owe asinii'" (2006). The English translation for this teaching is: The Anishinaabe uses tobacco when she or he goes to a sweat. This is the time the Anishinaabe is seen making an offering to our grandfathers, to these rocks. Here "asemaa" (tobacco) is treated as an animate being because the verb *"aabaji'"* (to use someone) is said in reference to asemaa. If asemaa were considered an inanimate object, then another verb, *"aabajitoon,"* would have to be used with it. Of further evidence, however, is the second sentence, in which Johnson says, "Anishinaabe owe biindaakoonaad owe gimishoomisinaanig owe asinii'." Here Johnson says that the Anishinaabe makes an offering to "gimishoomisinaanig" (our grandfathers). An animate verb is used in reference to "gimishoomisinaanig," indicating that the grandfathers are considered animate beings, just as they are in English. After grandfathers, Johnson says, "owe asinii'," indicating that the grandfathers of which he is speaking are rocks. These rocks are spoken of just as one would speak of human grandfathers or any other animate being.

### INAWENDIWIN: OUR RELATIONSHIPS WITH ALL OF CREATION

Within inaadiziwin, all of these animate beings are interconnected and dependent upon each other. Biskaabiiyang research terminology describes this interconnectedness as *inawendiwin*, referring to our relationships with all of Creation ("Anishinaabe Wordlist" 2003). There are dibaajimowinan explaining the interconnectedness and interdependency of every being. One of these, which I heard from Mary Geniusz many times while a child, Basil Johnston describes throughout *Ojibway Heritage* (1976). This dibaajimowin describes the levels of Creation. It explains that the first beings created on earth were the natural forces, including the rocks, the weather forces, and the aadizookaanag (those spirits who carry our ceremonies, teachings, songs, and stories). The second beings created were the trees and the plants. The third beings created were the *awesiinyag*, the nonhuman animals. The last beings created were the Anishinaabeg. When all of these beings were created, they promised *Gichi-manidoo* (the Great Spirit) that they would live together and help all the other levels of Creation survive. This was an important promise, for almost all of these beings need each other to survive. Without the plants, the animals would not be able to eat or breathe. Without the animals, the plants would not be able to move to new locations, and they would not have what we now call carbon dioxide. Without the plants and the animals, the Anishinaabeg would have no food or resources for their survival. Without the first level, including the sun, rain, and teachings, to guide them, all three of the other levels would perish. The only level of Creation that could survive without the other three is the first level, for they were here long before the other levels of Creation. However, if the rocks, natural forces, and aadizookaanag were here alone, they would not be content, for they would not have fulfilled their promise to Gichi-manidoo to ensure the survival of all levels of Creation.

Inaadiziwin places humans as an integral part of the cycle of life, rather than at the center of the universe or at some level above the animals, plants, and the rest of Creation. Joseph B. Casagrande describes the way that John Mink, a medicine man from the Lac Courte d'Oreilles reservation in Wisconsin, spoke of this part of inaadiziwin:

> For John Mink the line between the natural and the supernatural was thinly drawn. His world was filled with an infinite array of spirits and forces that could influence the affairs of men. Nor was man conceived as a creature apart from the rest of nature. For the Ojibwa, as for many hunting peoples, animals and men are akin and the differences between them lay chiefly in outward form. Animals are motivated as men are motivated, live in societies as men live, act as men act and their fates are intertwined. Thus, the Old Man told how when a bear was killed its four paws and head were placed in position on a rush mat and a feast given. The head was decorated with ribbons, beadwork or baby clothes and food and tobacco placed nearby; and speeches were made to the bear's spirit so that it would return to the village of bears and persuade other bears to allow themselves to be killed. (1955, 118)

Izhitwaawin dictates certain protocols, such as the bear ceremony Casagrande describes, which must be followed in order to keep the balance and maintain inawendiwin. If one performs this ceremony the bear will "return to the village of bears and persuade other bears to allow themselves to be killed." If one does not perform this ceremony, the bear may not do this, leaving the Anishinaabeg without many important resources, including the bear claws, skin, meat, and grease. Speaking of other protocols that must be followed to keep the balance, Keewaydinoquay says,

> Whenever it seems to human kind that it is necessary to change a balance for some desirable end, which the human believes is a good one, then we need to speak to these other beings. We need to talk to the rocks that are going to be ground up to make the foundation of our house and our walks. We need to explain to them why it is that we're asking them to change form. We need to have them understand and never take a plant for healing without first talking to the species and then to the particular plants that are going to be used, asking for their permission, and asking that they please give healing. (1991b)

These protocols—the bear ceremony described by Mink and the offering made before gathering described by Keewaydinoquay—are ways in which those following izhitwaawin show their respect for the rest of Creation and help to maintain the reciprocal relationships between humans and other beings.

Maintaining inawendiwin also requires that we do not waste the lives of other beings. One does not just kill a buck for a trophy set of antlers or a bear for an ornamental rug. If one takes the life of an animal, then one makes use of the usable parts of that animal. Kathryn Osogwin describes how the Anishinaabeg used to kill deer at the end of winter, around March. At this time, men wearing snowshoes would chase a deer, eventually clubbing it to death when it started to have trouble getting through the hard, icy crust that forms on the snow during that time of year. Osogwin says that the Anishinaabeg could kill many deer this way, but they "never killed a deer to be wasted." When they did kill a deer, they used all of the meat (EWV, notebook 20).

Plants and trees are an important part of inawendiwin, and the same protocols apply to them as apply to any other living beings. The question will inevitably arise here if, from the perspective of izhitwaawin, plants and trees are considered to be animate beings. Keewaydinoquay says they are, and she uses this concept as an example of the great differences between inaadiziwin and non-native philosophies (1991a; 1991b). People who work with Ojibwe language, however, tend to say that some plants are animate while others are inanimate. There really has not been enough research in this area to give a definite answer one way or the other, but it seems that a lot of the inanimate names refer not to the entire plant, but to a part of the plant. Sometimes the name for the seed, berry, nut, or fruit of a plant is an inanimate noun, while the name for the plant is an animate noun. For instance, *miinan* is one Ojibwe name for blueberries, but this word does not describe the whole plant, it only describes the berries. The name for the entire plant, *miinagaawanzh,* is an animate noun (Nichols and Nyholm 1995, 89). Similarly, *ozhaaboomin,* the berry of the gooseberry plant *(Ribes oxyacanthoides)* is inanimate, but the name of the plant itself, *ozhaaboominaganzh,* is animate (Rose, pers. comm.).[4] *Mitigomin,* the acorn of the white oak *(Quercus alba),* is also inanimate, while *mitigomizh,* the white oak tree, is animate (Rose, pers. comm.). Rose says that names for plants and trees that end in *-min* are referring not to the whole plant, but just to a part of the plant or tree, such

4. With the exception of *Sphagnum* spp., Rose identified all of the plants in this section from descriptions given in Meeker, Elias, and Heim (1993a). The scientific names are listed here as they are presented in that source.

as a nut or a berry. Many of these names ending in *-min* are inanimate. I do not have enough names for plants and trees to be certain if they are all considered animate, but of the ones I have collected from elders so far all are animate. The only exceptions I have found in my own collecting of plant names that are inanimate are ***doodooshaaboojiibik***, the dandelion *(Taraxacum officinale)*, and ***aasaakamig***, *Sphagnum* spp., the moss used for diapering babies (Rose, pers. comm.). These may be examples of inanimate plant names. Doodooshaaboojiibik, although a name for dandelion, refers specifically to the root of this plant because of the ending *-jiibik,* meaning root. If this name fits in the pattern described above, then there may be another name for this entire plant that is animate. As for aasaakamig, this name is usually translated as "moss used to diaper babies," and as such this name may be used in reference to the diapering material and not to the plant at all. It is also possible that within inaadiziwin, mosses are not considered plants.

## PROTOCOLS FOR GATHERING MATERIALS FROM PLANTS AND TREES

Despite these questions about whether or not all plants are considered animate, the teachings of izhitwaawin tell us that we must make an offering and ask permission before picking any botanical material. Whipple explains, "Akawe, akawe awiiya gegoo baa-mamooyan imaa akiing iw mashkiki, asemaa akawe giga-asaa ji-biindaakoodaman i'iw wegonen waa-mamooyan" (pers. comm.). Here she says that the first thing someone must do before going to pick medicines is to make an offering of tobacco. Rose explains that it is important to give asemaa when collecting plants, adding that if one is collecting many of the same kind of plant all at one time, this offering only needs to be made at the beginning of the collecting (pers. comm.). Keewaydinoquay explains that Nodjimahkwe taught her to treat plants just as one would treat an animal or another human being. When one wants to ask for the assistance of a plant, one makes an offering to that plant and asks that being for help: "One goes to the plant people and says, 'From the ancestors I have learned that your kind has healing,' whatever it is . . . 'and I ask you will you please heal,' and then . . . whoever it is, is actually recited to the plant like it was another being that you could communicate with. And then . . . you would promise that you wouldn't take so much of it that its grandchildren won't live after it, which means that you have to learn enough about that plant to learn how it reproduces so that you can see to it that that happens" (1989a). According to Keewaydinoquay, this offering is directed to the plant or tree being collected. She says that one must keep his or her promise to the plant if he or she wants the plant to keep its promise to heal. She says, "To our way of thinking, when you accept the gift of a plant you are, either by way of fueling up with plant food or by applying a plant medicine . . . asking that plant to become you" (1991a). It is important to ask a plant's permission before collecting it because if one does not ask that being's permission, he or she will not get that being's spiritual healing. Without the spiritual healing, a person may be physically cured but still sick because that person's spirit has not been healed. Keewaydinoquay says this often happens in non-native hospitals, when patients are released because they are physically healed—for example, their wounds are healed—but they are still sick (1991a). Hilger explains that the Anishinaabe consultant with whom she gathered plants also spoke directly to those plants when gathering their roots: "She dug a little hole with her hands, placed a small piece of plug tobacco in each, covered it with dirt, and spoke to the plant saying: 'I'll take just a little for my use, and here is some tobacco for you!'" ([1951] 1992, 92).

Other dibaajimowin say that this offering is made to beings who care for that plant. Huron Smith describes the Anishinaabeg with whom he worked placing "tobacco" in the hole from where a root was gathered. He explains that a song accompanies this offering and both of these things are given to "Grandmother Earth, to Winabojo, and to Dzhe Manido," so that they may make the medicine being prepared "potent" (1932, 349). When describing how John Mink instructed him to make medicines out of plants, Casagrande writes, "I was carefully admonished to place an offering of tobacco beside the tree or bush or in the root-hole of the plant where, he said, a blind toad always crouched" (1955, 114–15). Angeline Williams explains that if a plant is not properly respected, its owner will follow the one who did not respect the plant. She gives an example of a red flower that grows on the shores of Lake Michigan and is used as a medicine for children. Only adults are allowed to pick this flower, and they are only allowed one each. If an offering is not made when the flower is picked, the snake who owns that plant will follow the person who picked it. She explains that these snakes can "stand up tall" to see who picked these flowers. Williams tells a dibaajimowin about a woman whose daughters picked this flower and put it on their hats. A three-foot black snake with a white stripe and white head followed the girls, until their mother told them to put the flowers alongside the road. When they did this the flowers disappeared, and they assumed that the snake ate these flowers. Williams adds that another snake "guards" "swamproot,"[5] but that if this plant is gathered in the fall, the gatherer will not see the snake (EWV, notebook 19b).

Permission to pick some botanical materials requires more than an offering. Songs must be sung when gathering certain materials. Huron Smith explains that Nenabozho instructed the Anishinaabeg on what roots they should gather and what songs they should sing. Smith presents the following song as an example:

> Nin ba ba odji'bîke o'o'we'dasa'ssema
> *I go to gather roots; here is tobacco;*
>
> mînode ni nowi nîmîcîn gi wedji'bîkei'en
> *Give me direct guidance, you, maker of roots*
>
> da mino wi dji'bîkei'an.
> *That I may get the proper roots.* (1932, 344; italics in original)

The words of this song are similar to a root-gathering song that Keewaydinoquay taught to her *oshkaabewisag* (M. Geniusz, pers. comm.). From these lines, it appears that this song is to be sung when someone is going out to gather any roots, not one specific kind of root. It is also clear that the singer is making an offering to one who makes roots so that being will give the singer guidance. Densmore relates a story, which she received from an Anishinaabe man named Maiñ'gans,[6] about a song sung when preparing

5. Wheeler-Voegelin provides this name for "swamp root": *škigwusk škigwushk,* and she writes that it is "good for anything," such as cramps and convulsions. She adds that it is not poisonous (EWV, notebook 19b).

6. Spelled "Maingans" in a later publication (Densmore 1941, 550).

a medicine made from a specific root. According to this story, a man who was Midewiwin dreamed that water spirits came and told him how to make a medicine called *bi'jĭkiwûck'*. Maiñ'gans continues, "'In order to persuade them to return he composed and sang a song . . . He was a young man at the time, but he sang this song until he was old. He sang it whenever he dug the roots or prepared the bi'jĭkiwûck'. Others learned it from him and now it is always sung when this medicine is prepared'" (Densmore 1913, 63).

Izhitwaawin also requires certain ceremonies and feasts before or after gathering certain botanical materials. The protocols surrounding these ceremonies differ in different communities. Some of these ceremonies are open to the entire community, whereas others are reserved for those with a certain amount of training. Speaking of how the Anishinaabeg of long ago did things, Dora Dorothy Whipple says, "Akawe ge gii-ashangewaad igaye, zagaswewaad o'o gegoo ge wii-izhichigewaad, manoomin ige wii-mamoowaad oshki-miijiwaad. Noongom idash gaawiin izhichigesiiwag" (pers. comm.). Here Whipple says that the Anishinaabeg used to have feasts and ceremonies for what they wanted to do, or when beginning to harvest a food such as wild rice. Gene Prigge quotes James "Pipe" Mustache from Lac Courte d'Oreilles describing a feast held for maple sugaring: "'Maybe a week before sugaring we Indians would hold a feast and ask the Great Spirit for a blessing'" (1981, 1, 7). Ojibwe language teacher Dennis Jones (Pebaamibines) explains how one speaks in Ojibwe about giving a feast for a plant. He says that when someone is feasting a certain plant or tree, he or she says, "*Ingaagiizomaa*, 'I am feasting him or her,'" using the transitive animate verb *gaagiizom*. This verb is used when someone is talking about having a feast for a specific plant, but when someone is just having a feast in general, and not naming who that feast is for, that person would say "*niwiikonge*" (I am having a feast), using the intransitive animate verb *wiikonge* (pers. comm.).

### PERMISSION FOR GATHERING IS NOT ALWAYS GIVEN

It should be noted that just because one follows the proper protocols before gathering botanical material, one might not necessarily receive permission to take that material. Keewaydinoquay warns that if we use plants, or any other beings, for purposes to which they have not agreed, we will fail at what we are trying to do because that plant did not give its permission and is not lending its strength and spirit to our task. For instance, if a plant does not agree to be part of a medicine, that medicine will not cure its intended patient. As with any being, Keewaydinoquay says, there are ways to tell if a plant says yes or no when it has been asked if it may be used for a certain purpose: "Just as if you ask human beings . . . if you may walk through their property, there is something that indicates that it is all right with them. . . . Some people will care an awful lot and they'll say, 'Well you better not drop cigarettes . . . ' or 'Hell no! Can you see I've got a fence?' You can tell right away . . . what their attitude is going to be and whether they care . . . and you get the same kind of a reaction from the plants—the *whole* thing" (1991b). After giving this teaching, she tells her listeners about a time when one of her students made an offering to a plant and attempted to pick it. Before he could, a bee flew into one of the plant flowers and fell to the ground dead. The man went to Keewaydinoquay and asked her what she thought of this episode. She had him describe the plant to her, and then she told him that the plant was telling him that it was not all right for him to pick it because it was a poisonous plant. Keewaydinoquay tells another story about a time when, after making an offering, she went into a lake to pick cattails. The water started getting deeper and deeper, and the leaves began slipping out of her hands. She continues, "And after about five minutes I sort of caught on. I wasn't supposed to be collecting cattail plants. Although I had asked, the cattail plant had said, 'No.' It was a good idea that I got out of there because, by the time I got out, the waves were going right over the top of the cattails" (1991b). Keewaydinoquay freely shared information such as this with her students because this kind of gikendaasowin is important for anyone living inaadiziwin. As will be explained in the next section, there is other gikendaasowin that is guarded and not open to the public.

## GUARDED VERSUS PUBLIC GIKENDAASOWIN

Within izhitwaawin there are different kinds of gikendaasowin. There is gikendaasowin everyone must know for survival and to live according to the protocols of izhitwaawin, including the skills necessary to provide oneself with food, clothing, and shelter and the knowledge of how to properly ask permission before gathering botanical materials. Although many people today think of medicinal knowledge as "sacred" or "guarded," public gikendaasowin does include information about making simple remedies. Keewaydinoquay told Mary Geniusz many times that there was certain medicinal knowledge which, even as late as her childhood, was known by every member of the anishinaabe community in which she lived, and she openly taught this knowledge in many settings, including university courses and ethnobotany workshops (pers. comm.). Keewaydinoquay says that there are some medicines that "belong to the people." She instructs her oshkaabewisag that, although they may teach others how to make these medicines, they may not accept payment to do so (1990b). Whipple says that, although there was a medicine man or woman in every anishinaabe community, all the elders around her when she was a child made medicines (pers. comm.). Huron Smith also says this, explaining, "The patient usually calls the medicine man for ailments that have not responded to his own individual treatment" (1932, 351).

Along with this public gikendaasowin, there is also gikendaasowin that is more guarded, including Midewaajimowin (knowledge taught by the Midewiwin). This more guarded gikendaasowin is not commonly held by everyone, and is only given to specific individuals who have gone through certain ceremonies and degrees of training. For instance, while collecting information on anishinaabe plant use, both Densmore and Huron Smith discovered that they had to go to many different people to learn about various uses of plants because, although the people with whom they were working were Midewiwin, they did not all have knowledge about the same plants (Smith 1932, 345; Densmore ([1928] 1974, 322–23).

Those who have been entrusted with this more guarded knowledge often specialize in certain medicines and certain kinds of healing. Whipple describes the differences in the healing abilities of her cousin Jim Jackson, a ***mashkikiiwinini*** (medicine man), and her mother, Waaweeyakamig, also known as Emma or Ethel Mitchell, who was not a mashkikiiwikwe but was a member of the Midewiwin and could make medicines to treat her family. Whipple compares the medicine that her mother could make to aspirin, explaining that her mother could make a diarrhea medicine made out of chokecherries and an eye medicine out of roots. She says her mother knew what medicines to use to treat an illness, but that Jackson was a specialist, who could see inside of a person and determine what was wrong with him or her. Whipple describes Jackson by saying, "He was a real medicine man, a specialist," adding that his specialty was curing the lungs, which he did for Whipple a couple of times. Jackson also knew if a person was so sick that it was too late to help him or her (pers. comm.). Mashkikiiwinini Ron Geyshick describes his own healing specialty, which he received in 1982 during an illness. On the fourth day of this illness, "the Lord" came to Geyshick and gave him his healing powers. Geyshick explains that the Lord said to him, "'In order for you to believe in me, I am going to give you special powers. These are the power to heal, and an X-ray vision which will let you see the illness in a person.' I use this vision in my healing. It comes in four split-second flashes, as fast as I could flick my fingers at you" (1989, 23–24).

A mashkikiiwinini or a mashkikiiwikwe often guards his or her special medicinal knowledge, sometimes by disguising the ingredients in prepared medicines. Both Huron Smith and Densmore describe "medicine men" grinding medicinal ingredients so that they are unidentifiable to anyone else (H. Smith 1932, 35; Densmore [1928] 1974, 324–25). Densmore says that this was done purposely to prevent anyone from being able to duplicate these recipes. She explains, "medicine men frequently combined an aromatic herb with their medicines as a precaution against their identification" ([1928] 1974, 324–25).[7] Smith adds, "Even if one knew all of the ingredients, the amounts of each herb would be difficult to ascertain. Often, as in the case of Sweet Flag *(Acorus calamus)*, the amount must be very limited since the medicinal effect is so severe" (1932, 351). George McGeshick, Sr., says that when he and his family were living at Gete-gitigaaning, the Old Village at Lac Vieux Desert, the medicine man there would prepare medicines and give them to his patients without telling them the ingredients. He would give the patient more medicine when they needed it, but he would not tell them how to prepare it themselves (pers. comm.).

7. To this Densmore adds, "The fact that persons were willing to impart their knowledge of these ancient remedies for publication indicates that the attitude of the Chippewa toward their old customs is passing away" ([1928] 1974, 324–25).

## GIKENDAASOWIN COMES FROM THE MANIDOOG

Within izhitwaawin, gikendaasowin, both guarded and public, comes from many sources, including the *manidoog*, dreams, and animals. As stated previously, gikendaasowin is *gaa-izhi-zhawendaagoziyaang* (given to us in a loving way by the spirits). Some gikendaasowin is said to have come from various manidoog. Nenabozho, who is half-manidoo and half-human, brought Midewaajimowin to the Anishinaabeg. Johnston describes the origin of this information in his text. He tells about "Odaemin," a boy who died at a time when the Anishinaabeg were very sick. Odaemin wanted to do something for his people, and when "Kitche Manitou" heard this, he restored Odaemin to life, promising to send "Nanabush" to the Anishinaabeg so that they could learn about medicine. Nanabush comes to Odaemin and teaches him how to heal people with medicines. Johnston describes Nanabush's teachings: "Plants, he said, possessed two powers, the power to heal and the power to grow. Nanabush, moreover, taught young Odaemin that plant beings could lend their powers of healing and growing to other beings. Animal beings possessed this knowledge. Odaemin must learn from them. The greatest lesson that Nanabush imparted to Odaemin was how to learn." When Nanabush leaves, Odaemin continues to learn and observe the animals and how they use plants. He passes his gifts of healing on to another young man, who passed it on to others (1976, 80–82). Huron Smith writes a brief account, given to him by John Whitefeather, of how Nenabozho recreated the world after the flood and introduced plants to that world. He also writes of how Nenabozho brought the Midewiwin to the Anishinaabeg and taught them the proper protocols for gathering medicines (1932, 342–44). The aadizookaan "Nenabozho and the Animikiig," presented in chapter 4, describes how Nenabozho discovered and brought knowledge of *wiigwaasi-mitig* (the birch tree) to the Anishinaabeg. Geyshick says that healing powers, not just those related to plants, come from the manidoog. Specifically addressing young people, he writes, "Powers come from the spirits, not from humans. To learn, go out in the woods, and it may take days and nights of fasting for you to receive understanding" (1989, 32).

## GIKENDAASOWIN COMES FROM DREAMS

The manidoog often bring information to the Anishinaabeg in dreams, which are also considered a reliable source of gikendaasowin. Sometimes this information is given personally to the dreamer. I have heard members of my own family many times talk about curing themselves by following information they received in a dream. Kathryn Osogwin says that sometimes after their first menstruation, women have dreams about how to make medicines from certain plants, and later these women will see those plants and know how to use them (EWV, notebook 20). Angeline Williams explains that when an Anishinaabe is shown a certain plant in dreams by his or her guardian spirit, that person will try that plant when he or she awakes (EWV, notebook 19b). Anthropologist Paul Radin presents a story about an Anishinaabe man who received information this way. He was very sick but did not know how to make any medicines. By following the instructions he received in a dream about how to make medicine out of the "buttonwood" tree, this man recovered (1924, 523).

Sometimes the dreamer receives gikendaasowin to be shared with others. Williams describes a dibaajimowin that she heard from her grandfather about how a young girl first brought asemaa to the Anishinaabeg by following information she received in a dream. After being told that "a friend—a helpful person" would be in her bedding, the girl awoke and found a large plug of "tobacco" in her bed. For a long time this girl was the only one who had tobacco, and the other Anishinaabeg would buy pieces of it from her when they needed it to ask a powerful person questions or to have a feast. Every so often, the girl would find a new plug of tobacco in her bedding. Williams says that the Anishinaabeg knew how to use this tobacco because they received information about it in their dreams (EWV, notebook 19c).

Mashkikiiwikwe and mashkikiiwinini also receive instructions in dreams. I asked Whipple how someone becomes a mashkikiiwikwe or a mashkikiiwinini, and she said that a person has a dream or a vision when he or she goes on a fast, which tells that person what he or she is going to be. Whipple compares becoming a medicine person to being given a pipe, adding that these are gifts "from the high spirit or manidoo." She adds, "You can't just call anyone a medicine man" (pers. comm.). Smith writes that after checking a person's physical ailments, such as looking at his or her eyes or tongue, a medicine person dreams over the ailment: "Usually they want time to dream over the case, and drink a draught of their own dream-inducing medicine before going to sleep. In a vision or dream, they are directed to the proper medicine to use, and concoct it the following day" (H. Smith 1932, 350).

## GIKENDAASOWIN COMES FROM ANIMALS

Gikendaasowin also comes from animals, and sometimes the Ojibwe name for a plant or tree contains the name of the animal who brought information about that plant to the Anishinaabeg. For example, one of the Ojibwe names for jewelweed (*Impatiens capensis* Meerb.) is *omakakiibag* (frog leaf) (Zichmanis and Hodgins 1982, 263).[8] One dibaajimowin tells how an *omakakii* (frog) taught the Anishinaabeg about how to use this plant to heal skin irritations, such as the rash caused by poison ivy (*Rhus radicans* L.) and poison sumac (*Rhus vernix* L.). I have heard this story many times throughout the years from my mother, who learned it from Keewaydinoquay. Johnston refers to this story in *Ojibway Heritage* (1976, 42). In this dibaajimowin, the omakakii escapes being eaten by a snake by jumping into a patch of *maji-aniibiish* (poison ivy) and by waiting for the snake to go away.[9] Once the snake is gone, the omakakii jumps right from the patch of maji-aniibiish to a patch of omakakiibag. He rolls around in the omakakiibag, getting the juices of the plant all over his skin, and these juices stop his skin from being irritated by the maji-aniibiish. Johnston explains, "From the conduct of the little frog the Anishinabeg learned the cure for poison ivy" (1976, 42).

*Makwa* (bear) is an important healer who has also taught the Anishinaabeg about many plants. Members of the bear clan often become medicine

8. At the end of *Flowers of the Wild: Ontario and the Great Lakes Region,* Zichmanis and Hodgins provide a short list of Ojibwe plant names. There is no mention of the Anishinaabeg in the rest of their text. See references for full citation.

9. Biologist Dr. Peter Kaufman notes that poison sumac (*Rhus vernix* L.) grows in wet, swampy areas, often in the same places where jewelweed (*Impatiens capensis* Meerb.) grows, and that poison ivy (*Rhus radicans* L.) grows in more diverse locations and up the trunks of trees (Kaufman, pers. comm.).

people, as this is one of their roles within izhitwaawin. Keewaydinoquay says, "In our particular tribal group, Bear is considered the chief medicine spirit," and adds that medicine people often wear "bear objects" to honor makwa (1991a). She writes that Nodjimahkwe instructed her to always pay close attention to any plant named after makwa because these plants are always important, and for the rest of her life she should always try to learn all she could about any plant named after makwa. Reiterating Nodjimahkwe's instructions, Keewaydinoquay writes, "Now remember this assignment I have given to you; it is one to last for a lifetime so that the sources of Bear's healing shall not be lost to The People" (1977, 16–17). According to the aadizookaan "The Creation of *Nookomis Giizhik*," presented in chapter 4, Mishi-makwa and Nigig first brought *makwa-miskomin* (bearberry, *Arctostaphylos uva-ursi* L.) to the Anishinaabeg so that they would never be hungry. This plant, named after makwa, is a very strong medicine used in many ways, including the prevention and treatment of diabetes (Keewaydinoquay 1977, 21–22; 32–33).

## GIKENDAASOWIN DOES NOT COME FROM RANDOM EXPERIMENTATION

As seen from these examples, within izhitwaawin there are many sources of gikendaasowin. The Anishinaabeg do not go out into the woods and experiment with randomly selected plants and trees to see what can be made from each. Instead, they rely on generations of knowledge, originating from various sources. I know Anishinaabeg of my generation who believe that in times past our young people were sent out into the woods to randomly experiment with any plant or tree they could find, and that they would know, by the appearance of a plant, how that plant should be used. This is a very dangerous misinterpretation of the origins of botanical gikendaasowin. No one in a survival situation would send strong, healthy young adults out into the woods to taste, smoke, or rip up anything that they found because there are many poisonous plants and trees that those young people could encounter. The result of such random experimentation could cost a community entire generations of young people. The *Gete-anishinaabeg* (old-time Indians) were living among these plants and trees, and they must have known that there are certain plants and trees that will kill a human being. None of these stories suggests that they ever randomly experimented with plants. Instead, these ***aadizookaanan*** and dibaajimowinan tell us that the Anishinaabeg use information that they receive from some source, whether another human, an animal, a dream, Nenabozho, or a manidoo, to make the decision on how to use a previously unknown plant or tree. For example, Radin, H. Smith, and Wheeler-Voegelin all describe Anishinaabeg receiving information about how to make certain medicines in their dreams and then trying those medicines out when they awoke (Radin 1924, 523; H. Smith 1932, 350; EWV, notebooks 19a and 19c).

Mary Geniusz says that Keewaydinoquay was adamant that the "Doctrine of Signatures," the theory that one can tell by the appearance of a plant what part of the human body that plant can treat, was never a part of gikendaasowin. Keewaydinoquay insisted that a mashkikiiwinini or mashkikiiwikwe would quickly kill an oshkaabewis by instructing him or her to randomly experiment with plants because that oshkaabewis would inevitably encounter a poisonous plant. She often hypothesized that the Doctrine of Signatures was created in Europe after the people living there already had a large body of plant knowledge and were trying to figure out how it had originated (Geniusz, pers. comm.).

When experimentation is necessary, those living inaadiziwin rely on generations of knowledge to make informed decisions about which plant or tree they should use. The results of this informed experimentation is another source of gikendaasowin. Keewaydinoquay argues that mashkiki has such strength because it was gathered and contributed to over thousands of years (1989a). She explains, "And it's perfectly true that in times past, when there was some great scourge of disease, something like that, that a lot of elder people offer themselves to be experimented on. And this was not considered a bad thing. A lot of non-Indian people seeing that kind of thing said 'Oh isn't that awful. They tried these things out on their weak old people.' Well, see, the Indians didn't look at it that way . . . it was considered a worthwhile kind of thing to do" (1985). She says that when elders offered their lives to be experimented on like this, it was considered a very noble thing that they were doing for the good of the people. This sacrifice allowed the Anishinaabeg to know how these medicines worked on human beings, knowledge that, Keewaydinoquay argues, is more valuable than the non-native practice of experimenting on other mammals (1985).

## GIKENDAASOWIN IS MAINTAINED THROUGH STORIES

Gikendaasowin coming from these sources is maintained and used through a number of ways including stories, songs, oral teachings, formal apprenticeships, and personal notebooks. Dibaajimowin and aadizookaan are one method used by the Anishinaabeg to maintain and pass on gikendaasowin about plants and trees. Often gikendaasowin is found in an aadizookaan or a dibaajimowin explaining a plant origin or the origin of how the Anishinaabeg learned how to use a certain plant. The dibaajimowin about the plant omakakiibag, described earlier, explains how to use this plant to cure skin irritations. One simply, as the omakakii does in the dibaajimowin, rubs the juices of this plant against the irritated part of the skin to relieve the effects of poison ivy, poison sumac, and other minor irritations. The aadizookaan presented in chapter 4, "The Creation of Nookomis Giizhik," contains gikendaasowin describing the physical, medicinal, and spiritual uses of this tree. In a recording of this aadizookaan, Keewaydinoquay explains to her listeners that the part of the story where Mishi-makwa and Nigig are looking for a tree to use to make a hole in the earth is meant to be a teaching section. The first tree that they tried was "cotton wood aspen."[10] She continues, "Now cotton wood aspen is all right for beavers to chew, in fact the outer bark is their favorite and . . . it also contains good medicine, as we all know, but they put that aspen stick in the hole and they had only given two or three digs and it snapped right off because you know how aspen snaps." Then in an aside to her listeners, Keewaydinoquay says that, although she will not be doing so at this time, "One of the purposes of this story is to teach about the characteristics of all the different kinds of wood, and what they are good for, and what they aren't good for, and so they go through the whole list of all the different trees" (Keewaydinoquay n.d.*c*).

Other dibaajimowinan and aadizookaanan, which are not specifically focused on a particular plant or tree, can also contain gikendaasowin about plants and trees. For example, Frank Speck presents another teaching story

10. Probably a variety of *Populus deltoides.*

that he recorded while working with the Timiskaming Algonquin and Timagami Ojibwe. In this story, "Nenebuc" has his head stuck in a bear skull and is not able to see. As he walks along, Nenebuc has to use the trees to know where he is. Speck continues, "Then he felt the trees. He said to one, 'What are you?' It answered, 'Cedar.' He kept doing this with all the trees in order to keep his course. When he got too near the shore, he knew it by the kind of trees he met. So he kept on walking, and the only tree that did not answer promptly was the black spruce, and that said, 'I'm Se·'se·ga'ndαk' (black spruce). Then Nenebuc knew he was on low ground." This line of trees leads Nenebuc to a lake. After swimming across this lake and falling on a rock, Nenebuc cracks the bear skull off his head (Speck 1915, 33–34). In Speck's text, this story is quite short, only one paragraph long. The fact that two trees are presented here as key parts in the story so that Nenebuc knows where he is suggests that in a longer version more trees might be inserted, and this story could easily become a teaching story about the growing conditions of the various kinds of trees.

## GIKENDAASOWIN IS MAINTAINED THROUGH SONGS

Songs are also an integral part of teaching about gikendaasowin. Johnston writes, "Every plant had a place and a purpose; every plant had a time. For every plant being there was a prayer and a song" (1976, 83). Within izhitwaawin, there are songs about different plants and trees that contain dibaajimowinan honoring these beings. One example is a song that Keewaydinoquay called "Nookomis Giizhik: The Cedar Song," presented in chapter 4. There are lines in this song that clearly are meant to be a sign of respect for this tree, Nookomis Giizhik (Grandmother cedar, *Thuja occidentalis* L.). For example, the refrain of this song is: "Nookomis sa giizhik, gichitwaawendaagoziwin." *Nookomis* means "my grandmother" and *sa* is an emphatic particle. *Giizhik* is an Ojibwe name for the cedar tree. *Gichitwaawendaagoziwin* comes from *gichitwaawendaagozi*, an animate intransitive verb meaning "to be holy, revered, or venerable."[11] The ending *win* makes this verb a

11. This definition comes from Frederic Baraga, who writes: *Kitchitwâwendagos[i]* ([1878] 1992, pt. 2, 195).

noun. So "gichitwaawendaagoziwin" refers to an animate being who is holy, worthy of being venerated, honored, or revered.[12] This refrain translates as: "Grandmother! Cedar, one who is honored." There are also lines in this song containing gikendaasowin about this tree. For example, one line of this song says, "gaa-noojimowaad anishinaabeg." *Gaa-noojimowaad* comes from *noojimo*,[13] an intransitive animate verb meaning "one is cured, healed, or recovered." The conjugation *waad* means that "they" are completing this action. The prefix *Gaa-* is similar to the English relative pronoun *who*, but it can also indicate the past tense. So the implied translation of this line is: "the Anishinaabeg who are healed." This Ojibwe line corresponds to the English line of this song: "We call her saving tree, she saves the people." Both the Ojibwe and the English line refer to a dibaajimowin that says that this tree saves or cures the Anishinaabeg, physically and spiritually.[14]

Songs are also used within izhitwaawin as part of the healing process. Densmore explains that she was able to gather the information she used to write *Uses of Plants by the Chippewa Indians* because she was studying anishinaabe music. She says, "Songs were used by both the djaskid and Mide, and considered an essential part of their treatment. Having recorded these songs, the Chippewa were willing to impart information concerning the herbs and their uses" ([1929] 1979, 44–45).[15] Densmore presents two songs each labeled as a "Healing Song." She explains that these songs had been used to cure an elder woman named Mi'jakiya'cĭg (clearing sky) in her youth, and she shared them with Densmore (1910, 92–93).

Within izhitwaawin, the trees have another direct link to music: it is through their wood that the Anishinaabeg sing. As Mary Geniusz says, "You have to remember that we may sing about cedar but cedar also does our singing

12. Baraga includes this participle in his dictionary. He writes this participle *Kitchitwawendagosiwin*, and he translates it as "Honor, veneration, glory, glorification, sanctity" ([1878] 1992, pt. 2, 195).

13. *Noojimo* is a *vai* meaning "recover from an illness" (Nichols and Nyholm 1995, 102); Baraga has "recover," "cured," and "healed" as translations ([1878] 1992, pt. 2, 307).

14. This dibaajimowin comes from Keewaydinoquay, and it will be fully explained in chapter 4.

15. *Djaskid* is from *jaasakiid*: she or he who practices the shaking tent: a participle form of *jiisakii* (see glossary for more information).

for us" (pers. comm.). The *dewe'igan* (drum) and *bibigwan* (flute) are constructed from *giizhikaatig* (the northern white cedar) because this wood gets hollow as it decays (Keewaydinoquay 1986).[16] Densmore describes "courting flutes" being made out of this wood [1929] 1979, 12).[17] It is possible to find a cedar log in the woods that has already rotted enough to make a round for a dewe'igan. George McGeshick, Sr., describes searching in the woods for "ezhi-michaag giizhik, gaa-wiimbaakozid" (a hollow cedar log) to use to make a drum for the Chicaugon Chippewa Heritage Council's dances at Chicaugon Lake in the 1970s. Halfway between his home in Iron River, Michigan, and Watersmeet, Michigan, he found the stump of an old cedar tree that had been cut off high above the ground, possibly when deep snow covered the bottom half of the tree. He suspects this giizhikaatig (northern white cedar tree) was cut during the lumbering days before U.S. Highway 2 was constructed. McGeshick and those working with him cut a piece of this stump, and used it to make the dewe'gan who resides in their community (McGeshick, pers. comm.). There are many spiritual properties of giizhikaatig, as explained in chapter 4, and so musical instruments used for healing and other ceremonies are often made from this wood. Mary Geniusz says that Keewaydinoquay inherited a dewe'igan made of giizhikaatig from her grandfather, MidéOgema, and she used this dewe'igan for healing and other ceremonies (pers. comm.). Keewaydinoquay did not know the exact age of this dewe'igan, but he was well over one hundred years old when I saw him as a child in the 1980s.[18]

## GIKENDAASOWIN IS MAINTAINED THROUGH ORAL TEACHINGS

Gikendaasowin is also maintained through oral teachings, passed down generation to generation. H. Smith writes, "Individual knowledge was handed

16. It should be noted, however, that other trees are sometimes used as well. Densmore notes that the *Midewàkik* (the Midewiwin drum) that she saw is made out of basswood (1910, 12).

17. In her later text, Densmore notes that flutes are also made out of "box elder, ash, sumac, or other soft wood with a straight grain," although she does not identify these specifically as "courting" flutes ([1929] 1979, 167).

18. In izhitwaawin the drum is an animate being, often referred to as *nimishoomis* (my grandfather).

down through the family. Instruction to boys and girls usually comes from the uncle or aunt, and if they have no natural uncle or aunt, then one is assigned to them" (H. Smith 1932, 349). Whipple describes a similar descent of knowledge in her community, explaining that her mother learned about medicines from her mother, and that these teachings were passed on through the generations. Whipple also learned how to make some medicines from her mother, although she no longer remembers all of the ingredients. She says that her brother learned about medicines from both of his parents (Whipple, pers. comm.).

Being able to trace the descent of one's dibaajimowinan is an essential part of how gikendaasowin is maintained within izhitwaawin. Anishinaabe author Louise Erdrich gives an example of Tobasonakwut, a spiritual leader, using this protocol when sharing songs: "Tobasonakwut always begins his story of this song by attributing it to his uncle Kwekwekibiness. Very traditional people are very careful about attribution. When a story begins there is a prefacing history of that story's origin that is as complicated as the Modern Language Association guidelines to form in footnotes" (Erdrich 2003, 39). Densmore explains that this attribution is done with all songs: "The history of the Chippewa songs is well known to the singers, and is further preserved by the Indian custom of prefacing a song with a brief speech concerning it." Densmore adds that on formal occasions the singer of a song will introduce a song by saying whose song it is, and then they will conclude the song in the same manner (1910, 2). When making recordings of gikendaasowin with elders, I have found that they often explain the history of their teachings before giving them. For example, when explaining why it is important to make a tobacco offering during a thunderstorm, Whipple begins by explaining where she learned this dibaajimowin. She says, "Mii giiwenh mewinzha ko gaa-izhi-noondamaan, gichi-aya'aawiwag iw, ingii-pizindawaag niin ingiw ingichi-aya'aag. Idash ge niin sa go chi-aya'aawiyaan. Miish iw wenji-dibaajimoyaan i'iw. Geget igo, indebweyendam i'iw gaa-ikidowaad, ingii-pizindawaag niin ingiw ingichi-anishinaabemag" (Whipple 2006a). Here Whipple explains that this dibaajimowin comes from her elders, to whom she listened when she was younger. She adds that now she is an elder herself, and that this dibaajimowin is true.

Within izhitwaawin, the descent of one's dibaajimowin is so important that when presenting gikendaasowin, Anishinaabeg introduce themselves by giving the name of their clans, their home communities, and their ***anishinaabe-wiinzowinan*** (Indian names) so that those listening to them will know the origins of their teachings. Many of the elders whom Anishinaabe author Anton Treuer quotes in *Living Our Language: Ojibwe Tales and Oral Histories* begin sharing dibaajimowinan and aadizookaanan by first introducing themselves and their families. For example, Collins Oakgrove begins,

> Zhaawanoowinini indizhinikaaz, miinawaa dash a'aw ogiishkimanisii indoodem. Imaa wenjibaayaan, imaa Miskwaagamiwi-zaaga'iganiing, mii wenjiwaad ingitiziimag odoodeman migiziwan. Ganabaj a'aw nimishoomisiban Zhaaganaashiiwakiing gii-onjibaa. Gii-pi-izhaa omaa. Aabading gaa-ikidowaad ingitiziimag apane.
>
> My name is Zhaawanoowinini, and my clan is the Kingfisher. Where I am from, there at Red Lake, that's where my parents were from. And my late mother, she was of the Bald Eagle Clan. My grandfather may have been from Canada. He came here. One time he visited this woman there at Ponemah. That's what my parents always said. (Treuer 2001, 170, 171)

After giving this introduction of himself, his clan, his parents, and his grandparents, Collins Oakgrove introduces the aadizookaanan he is going to tell. Similarly, when Susan Jackson talks about her childhood, she begins by saying,

> Inger ingii-tazhi-ondaadiz, Chi-achaabaaning ezhinikaadeg. Mii iwidi nimaamaa, miinawaa nimbaabaa gii-ayaawaad. Mii iwidi ondakaaneziwaad, gaa-onji-gikendamaan akina gegoo gii-pizindawagwaa nimaamaa miinawaa nookomis, gaye gegoo gii-kagwejimagwaa gegoo waa-izhi-gikendamaan gii-izhichigeyaan gii-ani-mindidoyaan. Miish onow namanj gii-kikendamaan gegoo i'iw.
>
> I was born in Inger, Chi-achaabaaning as it's called. My mother and father were over there. That's where they come from, where I got my knowledge of everything from, listening to my mother and grandmother and asking them what I wanted to know in what I did as I got bigger. That must be how I learned these things. (Treuer 2001, 206, 207)

Here, after explaining a little about her family's background, Jackson says that she probably learned everything from listening to her mother and grandmother and asking them questions. She then shares some of the things that she learned from them (Treuer 2001, 206–9).

Dennis Jones says that his father told him that one always needs to introduce him or herself by stating his or her anishinaabe-wiinzowin, clan, and home community. Jones refers to this dibaajimowin as "Ojibwe protocol" and says that his father was instrumental in encouraging people in Canada Treaty Three Area to introduce themselves this way. Jones explains, "It is important to do in any introduction . . . public speaking and ceremonial speaking." Jones says that by stating their anishinaabe-wiinzowinan, speakers tell their audiences who they are and that their identity as Anishinaabe is important enough to them that they sought out an anishinaabe-wiinzowin. By stating their clans, speakers are telling their audiences to whom they are related. By stating where they are from, where their home communities are, speakers are telling their listeners the origin of their teachings (Dennis Jones, pers. comm.). Introductions such as the one Jones describes are an important part of izhitwaawin, and are used throughout anishinaabewakiing. Keewaydinoquay told Mary Geniusz that it was important for Anishinaabeg to introduce themselves this way, but she taught that it was important for speakers to always state their clan first, followed by their home community, and last of all their individual name. Speakers should put themselves last, thereby demonstrating their humility (M. Geniusz, pers. comm.).

Being able to trace the origin of one's dibaajimowinan is one part of the personal connection to knowledge that exists within izhitwaawin. According to Keewaydinoquay, one's dibaajimowin is not credible unless that person explains his or her own interactions with what is being presented. These interactions may include dibaajimowinan that come from the personal experience of the speaker or from the experience of someone very close to the speaker (Mary Geniusz, pers. comm.). Keewaydinoquay includes personal stories about her own experiences working with plants and trees in her publications. She begins the first chapter of *Puhpohwee for the People* with a personal account of how she and people who were very close to her worked with various kinds of fungi. In this chapter she describes watching MidéOgema, her grandfather, use the smoke from the burning of two kinds of puffballs, *Fomes igniarius* and *Fomes fomentarius,* to calm a hive of bees so that he could extract some honey from it (Keewaydinoquay 1998, 1–5). She begins another publication, *Min: Anishinaabe Ogimaawi-minan: Blueberry: First Fruit of the People,* with an account of how she first learned of the teachings of blueberries from an elder woman in her village named Nagowikwe (1978, i–iv).

Keewaydinoquay also includes personal stories about her own experiences working with plants and trees in her class lectures at the University of Wisconsin-Milwaukee. When lecturing on cattails (*Typha latifolia* L.), Keewaydinoquay begins by speaking about a time when she was a child and her mother hired a contractor to remodel a wash shed in their cabin and make it into a modern kitchen. When the contractor took down one of the walls in this shed, he found cattail down stuffed between layers of flattened cardboard boxes. Keewaydinoquay says that her grandfather must have put this cattail down there, when he built the wash shed, to insulate the wall. She uses this story as an example of how cattails are used as insulation, one of their many uses within izhitwaawin (1985). When lecturing about motherwort (*Leonurus cardiaca* L.), Keewaydinoquay talks about a young girl she had in her school who had a heart defect that caused her blood to not get enough oxygen. The doctors said she would not live past grade seven. The girl's family tried using Keewaydinoquay's method of treating the girl with motherwort, and Keewaydinoquay last saw the girl at her high school graduation (Keewaydinoquay n.d.*a*).

## GIKENDAASOWIN IS MAINTAINED THROUGH APPRENTICESHIPS

Formal training is also an important maintenance device of gikendaasowin. Often one who is very skilled in a certain area, such as making medicines or weaving, will take on oshkaabewisag (apprentices). Nodjimahkwe, a well-respected mashkikiiwikwe living in the village on Cat Head Bay, accepted Keewaydinoquay as an apprentice when the latter was approximately nine years old. Keewaydinoquay says her mother gave her a beautiful beaded bag and told her to go to Nodjimahkwe and ask her if she would teach Keewaydinoquay "something worth while for the people." She adds that in her community children were always given as an oshkaabewis to an older member of the community so that that child could learn certain skills. Keewaydinoquay says that Nodjimahkwe never withheld anything from her, and answered any question that Keewaydinoquay asked (1989a). As an oshkaabewis, Keewaydinoquay learned many things about gathering, preparing, and working with plants to make foods and medicines. She also learned how to work with a plant's physical and spiritual properties. Keewaydinoquay eventually trained her own oshkaabewisag, Mary Geniusz among them, who is now training me.

Written sources also mention a system of formal training within anishinaabe communities. Huron Smith writes, "The young man who had the proper dream following the period of fasting in his youth, predicting his predilection towards the medicine man profession, was taken through a rigorous course of training" (H. Smith 1932, 349).[19] Anishinaabe James "Pipe" Mustache says his training to become a mashkikiiwinini began when he was "'about 15 or 18.'" He says at this time the elders saw that he was interested in religion and started to teach him many ceremonies, including ones for births, deaths, and marriages. They did this by bringing Mustache to ceremonies so he could "copy their methods" (Prigge 1981, 1, 7).

## GIKENDAASOWIN IS MAINTAINED THROUGH PERSONAL NOTEBOOKS

Some gikendaasowin is kept in personal notebooks and on scrolls made of *wiigwaas* (birch bark). Mary Geniusz says that Keewaydinoquay had at least two bound notebooks in which she kept gikendaasowin. One looked much older than the other. As part of Geniusz's oshkaabewis training, Keewaydinoquay taught her medicine songs and then let her copy the song from her newer notebook. Geniusz says that because the protocols of izhitwaawin prohibit an oshkaabewis from randomly paging through a personal notebook of his or her teacher, she only saw the portions of the notebook that

19. It should be noted in this passage that Huron Smith talks mainly about the practices of men because it was mainly men, he says, who were willing to talk to him about plants. Take, for example, the following statement by Smith: "Most of our informants were men, because they found it easier to talk to the writer than the women. It was easy to get the women to talk of old time methods of preparing aboriginal foods" (1932, 334).

Keewaydinoquay wanted her to see. These selected portions contained songs used for certain ceremonies. Keewaydinoquay told Geniusz that Nodjimahkwe kept written records of medicinal recipes on a writing tablet and healing songs on wiigwaas scrolls (Geniusz, pers. comm.).

Evidence of these personal notebooks exists in the written literature as well. Albert Reagan says that the information found in some of his articles was copied from the personal notebooks and wiigwaas scrolls belonging to medicine men at Nett Lake. As described in chapter 1, Albert Reagan published two articles presenting medicinal songs and recipes from George Farmer's notebook (1921; 1922). Reagan also says that the information in his article "Plants Used by the Bois Fort Chippewa (Ojibwa) Indians of Minnesota" comes from a "direct translation" of the information he found "in a medicine man's notebook" (1928, 231). Because the recipes given in this article are different from those in Reagan's article on George Farmer's medicinal recipes, one can assume that Reagan has seen more than one of these notebooks. Reagan also describes medicine men at Nett Lake singing medicine songs recorded in "picture writing" and written on "birch bark parchments." One mashkikiiwinini, he says, had over forty of these parchments (1927, 81–83).

These notebooks are designed for the personal use of their author, so each one differs in content and presentation of information. Reagan says that Farmer's notebook is written completely in Ojibwe using the English alphabet, and he claims that he copied the Ojibwe lines in his articles directly from this notebook (1921, 246; 1922, 332). Keewaydinoquay told Geniusz that some of the recipes and songs in Nodjimahkwe's notebooks were recorded in "hieroglyphs." The songs that Geniusz was allowed to copy from Keewaydinoquay's notebook were written mainly in Ojibwe using the English alphabet. Some passages were recorded in "hieroglyphs." Geniusz's own notebook contains copies of these songs, written in Ojibwe using English letters. Geniusz also uses English in her notebook to remind herself of instructions to be followed during certain ceremonies and to translate some of the Ojibwe songs (M. Geniusz, pers. comm.).

These notebooks are personal mnemonic devices for the people who create them, not a means of leaving a written record for future generations. Keewaydinoquay, for example, instructed her oshkaabewisag to bury her with her notebook. Geniusz says that is often done with notebooks so that a mashkikiiwikwe or a mashkikiiwinini will not leave his or her oshkaabewisag with information that they have not studied in the proper manner. It would be very dangerous for someone without enough training to try out a medicine just from reading about it in a notebook. For one thing, Geniusz explains, these notebooks do not always contain all of the information necessary to conduct a certain ceremony or to make a certain medicine. Key ingredients are often left out of these notebooks so that the person writing down the information could control who had it and who did not (M. Geniusz, pers. comm.). We can see examples of these abbreviated medicinal recipes in the lines that Reagan copied from Farmer's notebook, such as this recipe used to treat gonorrhea:

> (*a*) Adjimag, (*b*) mitigomish, (*c*) anib, (*d*) shishi-gi-me-wish, (*e*) asa edema
> ash oak white elm sugar maple put in tobacco
>
> we-da-bag dji-ga-tig ko-ko-sa-wet (or ho-ko-sa-wet)
> east little close trees gonorrhea (1921, 247)

Reagan explains that he added an English translation and an explanation to each recipe for presentation in the article. The rest he copied entirely from Farmer's notebook (1921, 246). Looking at the original Ojibwe recipe, then, when compared to Reagan's explanation, it is clear that Farmer has left some important information out of the recipe as recorded in his notebook. For instance, he does not say what part of these trees to use, or how to prepare and administer this medicine. These things are explained by Reagan: "For gonorrhea make a tea of the root-bark of the following trees: ash, oak, white elm, and sugar maple; add a little tobacco and set the solution just east of and quite close to some trees. When it is cool drink a cupful three times a day" (1921, 247–48). Farmer did not need these specific instructions, as he undoubtedly knew how to make the recipes recorded in his notebook. Anyone else, however, would need more information to reproduce these recipes.

Songs kept in these notebooks and on these scrolls of wiigwaas are also kept in a way that prevents them from being sung by anyone. Even if one can read the method through which a song was recorded—"pictographs" or Ojibwe with English letters—that person will still not necessarily be able to sing that song because these written songs do not provide the tune to which the song is sung. Those songs that Reagan says he copied directly from Farmer's notebook are just lines of Ojibwe, with no music notations (1922, 333–65). I have seen the songs in Mary Geniusz's notebook, and they also have no indication of the tune to which they are sung. If one happened to know the tune to which a song was sung, that person would not necessarily know other crucial information about the song, such as when the song should be sung and how the song should be used.

## GIKENDAASOWIN IS MAINTAINED THROUGH A RECORDING SYSTEM

Some Anishinaabeg also maintain gikendaasowin through symbols, often referred to as "pictographs" or "picture writings." There exists a large body of literature on anishinaabe "pictographs" and those of other tribes, going back at least as far as the research of Henry R. Schoolcraft (1851–57) and Garrick Mallery (1886, 1893). Specific colonized presentations of these symbols are discussed in chapter 3. Although non-native scholars argue that these "pictographs" are not a "writing system" (Gelb 1952, 35–36; Rogers 2005, 2–3) in which certain symbols stand for certain sounds, they are a means of recording messages, thoughts, ideas, and information. Keewaydinoquay explains that this system records thoughts rather than exact words, but it can be used to depict anything, although some ideas are harder to depict than others (1989a; 1990a). Mary Geniusz says that from what Keewaydinoquay taught her of this system, there are some concepts that are codified and some that are not. She says that a person learns a basic set of vocabulary, upon which that person is expected to elaborate should it be insufficient to express a certain piece of information (pers. comm.). For example, in Keewaydinoquay's system one places two circles around a group of hieroglyphs to indicate that they represent a name (1990a). I often saw Keewaydinoquay using this system of recording ideas, and she gave my mother several small wooden plaques on which she had painted various teachings using this system. When she named my sister and me, Keewaydinoquay taught us both to draw our names using this recording method.

In the written literature, there exist several descriptions of Anishinaabeg using a system of recording thoughts, ideas, and messages that seems to be similar to that which Keewaydinoquay used. As previously described, Reagan discusses songs written in "picture writing" on birch bark. Walter James Hoffman describes a "pictorial résumé of the traditions and history of the Ojibwa cosmogony" used by the Midewiwin at Red Lake, approximately fifteen feet long and twenty inches wide, made of pieces of birch bark sewn together (1888, 218). Densmore says that the Anishinaabeg use "pictography" on birch bark, cedar, ash, other woods, the "perpendicular surface of rocks," dry earth, and the dry ashes from fires. When they use pictography on birch bark, Densmore says that the Anishinaabeg generally use only the inner bark, but that the Midewiwin scrolls often make use of both sides of one piece of bark ([1929] 1979, 174). M. Inez Hilger describes members of the Midewiwin singing songs at a ceremony that were recorded in "pictographs on birchbark rolls" ([1951] 1992, 67).

Although not exclusively used to record botanical gikendaasowin, some of this knowledge is recorded using this system. Keewaydinoquay told Mary Geniusz that Nodjimahkwe taught her to read this writing system by reading out of her personal notebooks. Keewaydinoquay taught Geniusz how to read this system by letting her read the "hieroglyphs" written in her notebook (M. Geniusz, pers. comm.). She also uses it in *Min: Anishinaabe Ogimaawi-minan: BlueBerry: First Fruit of the People* to describe a dibaajimowin about blueberries (see illustration 1). This dibaajimowin encourages families to dry many blueberries during the Blueberry Moon so that they will keep their strength through the winter (Keewaydinoquay 1978, 33). It is obviously public information used to teach all Anishinaabeg an important survival skill.

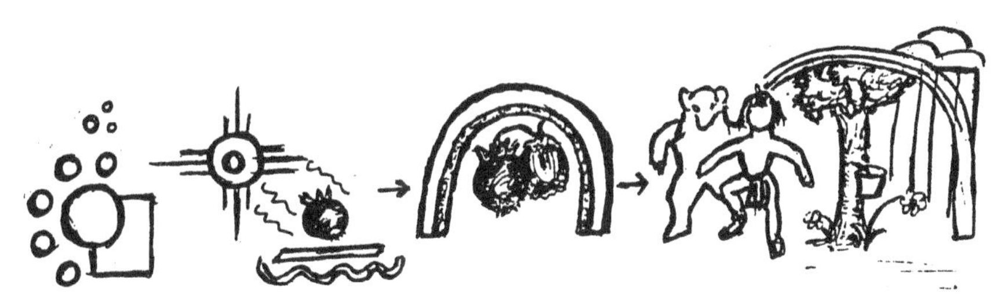

1. Ojibwe "hieroglyphs" as drawn by Keewaydinoquay (1978, 33). Courtesy of Miniss Kitigan Drum.

Other researchers presenting recording systems do not specifically describe them as being used to record medicinal recipes, but they do describe them as being used to record songs, including ceremonial songs, that may have been used in conjunction with certain medicines. These researchers also say that this system is used to record different kinds of gikendaasowin, some of which is guarded and some of which is not. Reagan says this "picture writing" is used to record "winter counts, medicinal formulas and songs, tribal history and lodge rituals," to be "inscribed and handed down from generation to generation" (1927, 81). Hoffman says that the Midewiwin are experts in the use of "pictorial writing," but it is not a tool used exclusively by them. He gives an example of a love letter written in "pictographs" by a woman to her sweetheart asking him to call on her (Hoffman 1888, 223). Densmore says that "pictography" is used to confer different kinds of knowledge: guarded knowledge, which is understood only by those in the Midewiwin, such as Midewiwin song records, writings, and stories about Nenabozho; and public knowledge, such as names, clans, maps, story illustrations, and messages left by travelers. Densmore claims that the kinds of "pictography" used to pass on the messages for the public have a "simpler symbolism" than those used to record gikendaasowin held by the Midewiwin ([1929] 1979, 174–75). Hilger says that of the Midewiwin songs she saw recorded in "pictographs," only the members of that organization could understand them ([1951] 1992, 67).

Although having varying degrees of details in the figures and other symbols, the recording systems described and exemplified by these other researchers sound and look like the one Keewaydinoquay used, but it is possible that each community has its own set of conventions for using this system, which may have developed independently of each other. For example, Keewaydinoquay says that her grandfather, MidéOgema, taught her a certain system for depicting if the event being recorded happened to a man or a woman. Keewaydinoquay adds, however, that she has not seen other people make this distinction in the same way that her grandfather did (1989b). Mary Geniusz says that this system is taught from one person to another, often from a mashkikiiwinini or a mashkikiiwikwe to his or her oshkaabewisag, and because it is handed down in this manner not everyone writes exactly the same. Geniusz knows how to read Keewaydinoquay's system because Keewaydinoquay taught it to her, but she does not think that she would be able to fully read and understand the writing system of another mashkikiiwinini or mashkikiiwikwe who was taught by a different set of teachers. Geniusz says that some medicine people would not necessarily want individuals who were not their oshkaabewisag to be able to read his or her dibaajimowinan written this way because then that person would lose control of those dibaajimowinan (pers. comm.), and as will be explained in the next section, proper payment must be secured before certain dibaajimowin and other kinds of gikendaasowin can be shared.

## SOME GIKENDAASOWIN MUST BE PURCHASED

When someone is entrusted with certain gikendaasowin, it is important that proper payment be made for that knowledge. Within izhitwaawin, there is always a price to be paid for gikendaasowin. Densmore says that members of the Midewiwin learn certain medicines when entering the Midewiwin and when advancing to different degrees, but any other medicines they want to learn have to be purchased from the "old men" ([1928] 1974, 322–23). Later Densmore says that in the "old days" nobody gave any medicinal information to anyone, not even family members, without sufficient payment. She explains, "one reason for this restriction seeming to be a fear that the information would not be treated with proper respect" ([1928] 1974, 324–25). Geniusz agrees, saying that it is important to charge for gikendaasowin so that those receiving it will respect and value it (pers. comm.). Ron Geyshick advises young people, "Help old people, and pay for the knowledge and help they give you. If you have a lot of food, share it. What you give you'll receive back" (1989, 31). Angeline Williams says that one must believe in a medicine and pay for it in order for that medicine to be effective. She adds that if one believes in a medicine but does not pay for it, that medicine will not do that person much good. Williams also tells a short story about a man who unknowingly purchased some fake medicine from another man. Later, the man who sold the medicine saw his customer in town, and the man was very grateful for the medicine that he had purchased. He said the medicine had worked wonderfully. Williams adds to this that "Indian medicine" works, whether it is fake or not,[20] as long as it is purchased (EWV, notebook 19b). Kathryn Osogwin says that curative medicines are paid for with a variety of things, including necklaces, pillowcases, or silk sashes. She says that some medicines are worth more than others. She briefly describes her father's medicine for stopping a hemorrhage as an example of a valuable medicine (EWV, notebook 20). Geniusz says that an oshkaabewis pays with "time, energy and support" for the information that he or she receives from his or her mashkikiiwinini or mashkikiiwikwe. Geniusz adds that this is how she paid Keewaydinoquay for the gikendaasowin that she was taught (pers. comm.). Mashkikiiwinini John Mink learned a lot of his "pharmaceutical knowledge" from his older relatives, but he also paid "substantial fees to other native doctors" for some of his medicines. He learned about other medicines in his Midewiwin training (Casagrande 1955, 114). Huron Smith says sometimes patients can buy instructions for making medicines from those "medicine men" doctoring them, adding, "The recipient is admonished to see that he does not impart the knowledge unless he is well paid for it, as he paid the medicine man" (1932, 351).[21]

Another form of purchasing knowledge is trading. In some cases gikendaasowin is exchanged piece for piece, as is often done with Midewaajimowin and other guarded knowledge. Mary Geniusz says this trading is done between peers, those who are at the same degree in the Midewiwin. Someone who is at a higher level of the Midewiwin generally does not trade Midewaajimowin with individuals who have not had as much training because the person with less training could easily hurt him or herself by not having the expertise to properly use that knowledge (M. Geniusz, pers. comm.).

20. In the actual notebook, Wheeler-Voegelin abbreviates these words as "Ind. med." (EWV, notebook 19b).

21. To this Smith adds, "This explains the difficulty one encounters when he tries to get medicinal information. Only by completely securing the confidence of Indians, can a white man get this information without pay, and then it must be thoroughly understood that the investigator is not copying their medicines to take commercial advantage of this knowledge. The Indian is quick to appreciate favors and to acknowledge the respect that is given to him by the white man, and becomes quite confident when he realizes that his confidence is not abused (H. Smith 1932, 351).

Keewaydinoquay says that her oshkaabewisag are rewarded in their third, fourth, and fifth years of service with new medicinal formulas, and once an oshkaabewis is gifted with one of these formulas, it is his or hers to barter or sell as he or she chooses (1990b).

## CONCLUSION

As seen in this chapter, those practicing izhitwaawin gather and maintain knowledge in ways different from those following non-native teachings and philosophies. Within izhitwaawin, plants and trees traverse physical and spiritual worlds. Within non-native teachings and philosophies, plants and trees are objects to be studied. There are also similarities between these knowledge-keeping systems, but understanding the differences between how izhitwaawin and non-native philosophies and teachings view plants and trees is the first step for those trying to decolonize this knowledge. Accepting these differences and seeing plants and trees through the perspective of our ancestors is the next step.

# 3

# The Colonization and Decolonization of Anishinaabe-gikendaasowin

## INTRODUCTION

Taking botanical information out of colonized texts and working with elders to make it a viable part of our communities would be useful to programs and individuals revitalizing izhitwaawin. However, as mentioned previously, the colonization process has reached far deeper than merely moving information out of one knowledge-keeping system to another and then presenting it to the world. Colonization, and the systemic racism it supported, was and is about one people completely absorbing another. To be truly effective, colonization must take over every aspect of that "other people," even the way in which they view the world. This chapter describes the colonization and decolonization processes, particularly as they relate to botanical gikendaasowin.

## COLONIZATION

The concept that colonization has reached into the minds, knowledge, and beings of indigenous people may be the hardest facet of colonization for us to accept and reverse. One can see that our lands and natural resources have been colonized just by looking at them: we no longer control, and in many cases no longer have access to, much of our ancestral lands or the resources on those lands. We can also see how our lives have been colonized: many of us now live on reservations or in places dictated by the dominant society and by our social and economic standing within that society. Some of our bodies have also been colonized in death, by continuing to linger on shelves and storage facilities of museums, universities, and other colonizing institutions. Unlike these examples, the colonization of our minds is internal and therefore not as readily seen at first glance. Some of us might not want to admit that this internal colonization exists because admitting that colonization is inside us makes us feel like the enemy.

Using a variety of assimilation efforts, colonizers have attempted to form indigenous peoples into their own image. It is not just a matter of taking indigenous children away to boarding schools and teaching them reading, writing, and arithmetic. Nor is it just a matter of breaking up tribes and putting individual families onto allotments, or relocating them into cities. Rather, it is a case of trying to assimilate indigenous people so that they will see the world and themselves from the perspective of the colonizer.

In *The Colonizer and the Colonized,* Tunisian theorist Albert Memmi describes how indigenous peoples are colonized internally. Memmi argues that the colonizer ascribes certain traits to the colonized, and by insisting that these traits exist, the colonizer can justify his actions toward the colonized. Memmi uses as an example the colonizers describing the colonized as "a wicked, backward person with evil, thievish, somewhat sadistic instincts." Once they establish such an image, the colonizers then claim that they need to treat the colonized harshly to protect themselves from these "evil" beings. In treating them harshly, the colonizers further argue, they are protecting the colonized from harming themselves. This image created by the colonizer is completely negative: "The colonized is not this, is not that. He is never considered in a positive light; or if he is, the quality which is conceded is the result of a psychological or ethical failing" (Memmi 1965, 83–84). The colonizer ascribes more and more traits to the colonized until, Memmi argues, "all the qualities which make a man of the colonized crumble away." As the colonizer continues to treat the colonized according to the image that he has created, the colonized begins to conform to the colonizer's image. He undergoes a change, becoming "a pure colonized." The colonized begins to see himself as this image created by the colonizer. He starts to believe it. Memmi concludes, "In order for the colonizer to be the complete master, it is not enough for him to be so in actual fact, but he must also believe in its legitimacy. In order for that legitimacy to be complete, it is not enough for the colonized to be a slave, he must also accept this role" (1965, 79–89).

During the assimilation process, indigenous peoples are taught to view themselves, their peoples, and their cultures from the perspective of the colonizer. The view that "a pure colonized" has of him or herself is negative, degrading, and of something not fully human. Linda Tuhiwai Smith describes one facet of how indigenous peoples have been viewed by their colonizers: "One of the supposed characteristics of primitive peoples was that we could not use our minds or intellects. We could not invent things, we could not create institutions or history, we could not imagine, we could not produce anything of value, we did not know how to use land and other resources from the natural world, we did not practice the 'arts' of civilization. By lacking such virtues, we disqualified ourselves, not just from civilization but also from humanity itself. In other words, we were not 'fully human'; some of us were not even considered partially human" (L. Smith 1999, 25). As Smith explains in this quote, from the worldview of the colonizer, indigenous peoples are "primitive," incapable of producing, doing, or thinking anything of value. They are not only uncivilized; they are not human. When indigenous peoples are assimilated and accept the worldview of the colonizer, this is how they see themselves, their knowledge, and their cultures. Memmi argues that the colonizers purposely force the colonized to make this transformation in order to reach their own objectives. If colonized people are made to think of themselves as less human, less intelligent, and less able than the colonizers, then they are easier to subjugate. They are not as tempted to rebel against the colonizers; they will put up with whatever the colonizer forces them to do (Memmi 1965, 79–80). This gives the colonizers incredible power over the colonized, their lands, their knowledge, and their resources.

For indigenous individuals who have been colonized, the thought that their people's language and knowledge could produce anything of equal or greater value than those produced by non-natives seems preposterous. For the colonizer's view of indigenous cultures is of something "primitive," "simplistic," and far inferior to the great knowledge and breakthroughs made by non-native civilizations. This belief works to the benefit of the colonizers, for when the colonized are made to see that their languages and knowledge are of no value, they are far more willing to part with those things. They stop using them. They forget them. They do not teach them to their children. Without languages and knowledge that are different from those of the colonizers, indigenous peoples are easier to assimilate and absorb within the dominant society. Once they are members of the dominant society, indigenous peoples are less likely to see themselves as distinct peoples and less likely to fight for their rights as sovereign nations.

Linda Smith argues that knowledge and cultures of indigenous peoples were specifically targeted by colonialist powers. She writes, "Knowledge was also there to be discovered, extracted, appropriated and distributed. Processes for enabling these things to occur became organized and systematic" (1999, 58). Smith says that various academic disciplines have collected indigenous knowledge. She adds, "It is through these disciplines that the indigenous world has been *re*presented to the West and it is through these disciplines that indigenous peoples often *re*search for the fragments of ourselves which were taken, catalogued, studied and stored" (1999, 58–59). Gikendaasowin has been collected and *re*presented to the world for more than 170 years. Many of these publications have negative passages about the Anishinaabeg as well as about their knowledge. The gikendaasowin in these colonized texts has been transformed to fit into non-native knowledge-keeping systems. Today Anishinaabeg are studying the colonized representations of gikendaasowin to learn more about their own culture, but such colonized writings are poor tools for cultural revitalization. They need to be decolonized so they can be made useful to the revitalization of izhitwaawin. What follows is a description of some of the colonized aspects of these texts on botanical gikendaasowin.

## COLONIZATION AND TEXTS OF GIKENDAASOWIN

The colonizer's invented negative views of indigenous peoples as "primitive," not capable of original thinking, run through many of the published presentations of botanical gikendaasowin. For instance, when introducing a list of plants used by the Anishinaabeg, Walter James Hoffman writes, "It is interesting to note in this list the number of infusions and decoctions that are, from a medical scientific standpoint, specific remedies for the complaints for which they are recommended. It is probable that the long continued intercourse between the Ojibwa and the Catholic Fathers, who were tolerably well versed in the ruder forms of medication, had much to do with improving an older and purely aboriginal form of practicing medical magic" (1891, 197). In this passage, Hoffman suggests that the Anishinaabeg are not intelligent or advanced enough to have gathered and preserved this knowledge without the help of non-natives.

The labeling of the anishinaabe method of recording thoughts and knowledge, including botanical gikendaasowin, as "pictographs" or "picture writing" is another area in which we are depicted as "primitive" people. Researchers argue that our method is not a "writing system" because it does not correspond directly to sounds in the Ojibwe language. For instance, linguist Henry Rogers says that writing "has a systematic relationship to language, and it has a systematic internal organization of its own" (2005, 3). In this method of recording information, symbols respond to ideas, not specific words or sounds, so perhaps this is not a "writing system" given this definition. That said, the ways in which this method of recording information are labeled by various researchers suggest that the Anishinaabeg simply have not reached a high enough rung on the ladder of evolution. For example, in "Interpreting Birch Bark Scrolls," Joan M. Vastokas describes this system as the "first step in the evolution of writing" (1984, 431). Here Vastokas suggests that the Anishinaabeg are at some beginning stage of development in their intellectual capacity. They are at the "first step" in creating a writing system. In a more recent text, Rogers describes this system as "non-linguistic graphic communication," and labels it "picture writing" (2005, 3). Others (Reagan 1922; Hilger [1951] 1992) use the term "pictograph," which is just a fancier sounding synonym. "Picture writing" sounds like something a child or an absent-minded person doodling on an envelope would do. Some researchers argue that labels such as "pictograph" and "picture writing" are not primitive enough. Linguist I. J. Gelb argues that the term "pictograph" is too advanced a label for an American Indian method of recording information: "'pictographic,' meaning 'picture writing,' is not appropriate because there are other systems, such as Egyptian, early Sumerian, etc., also expressed in picture form, but entirely different in inner structure from such primitive systems as those used by the American Indians" (1952, 35). Although coming from a text now over fifty years old, this quote does give readers a sense of the ways in which this method of recording information has been described and labeled by the colonizers. Overcoming such views is still an issue for those working with the decolonization of gikendaasowin and other native people's knowledge.

If picture writing is the first step of writing, one might assume that if colonization had not happened, eventually the Anishinaabeg would have developed a "true" writing system. Such assumptions ignore the possibility that oral methods of communicating and maintaining knowledge can be just as advanced as written ones. I have worked with several elders who cringe at the idea of writing any Ojibwe word. They say that our language was supposed to be spoken, not written. Not having a means of representing every sound in their language does not make a people "primitive." It is simply a different means of maintaining knowledge.

Keewaydinoquay resents simplistic and primitive sounding labels such as "picture writing" and "mnemonics," often used in reference to this system. She prefers to call these symbols "hieroglyphs": "Different people have called hieroglyphs many different things. We are talking about a way of marking that indigenous people once used to help them remember things. Because it is simple, most people in these days refer to it as 'picture writing,' 'mnemonics,' they hardly ever dignify it as a hieroglyph. And yet, it's enough like the hieroglyphs and has changed over the amount of years it has been used as the hieroglyphs have" (Keewaydinoquay 1990a). She adds that when her father was in the Spanish American War in the Philippines, he could read the hieroglyphs that he saw there because of his knowledge of anishinaabe "hieroglyphs" (1990a). Whether or not one agrees with Keewaydinoquay's label of this recording system, she does accurately describe the way in which it has been depicted by non-native researchers. Just because the Anishinaabeg did not, before the Invasion, design a "writing system" like those of non-native societies does not make us "primitive" or our method of recording information somehow inferior to non-native ones. If this method of recording was created entirely anew by each person who used it, then others would not be able to interpret it. As mentioned in chapter 2, there are, at least in individual regions, standard methods of recording certain pieces of information with this system. It is a method of recording information, but it is different from those used by non-natives. Non-native researchers' inability to interpret these records without the help of Anishinaabe consultants should prove that this is a system of record keeping that one must learn, rather than a simplistic, primitive means of recording information. If it was simplistic, everyone would be able to read it.

The negation of indigenous histories can also be seen repeatedly through the colonized presentation of gikendaasowin. Linda Smith describes the feelings of having one's history put down by others, saying, "There are numerous oral stories which tell of what it means, what it feels like, to be present while your history is erased before your eyes, dismissed as irrelevant, ignored or rendered as the lunatic ravings of drunken old people." She argues that the negation of indigenous histories has been an important part of colonization because these histories were "regarded as 'primitive' and 'incorrect' and mostly because they challenged and resisted the mission of colonization" (1999, 29). Anishinaabe histories about the origins of certain plants are generally not included in these colonized texts of gikendaasowin. If they are included, they are often dismissed as fictional stories of a simplistic, "primitive" people. For example, Huron Smith writes,

> The Ojibwe still use the songs essential to digging medicine roots. Jack Doud, the old scout captain of the Civil War, of the old Flambeau village, told the writer that Winabojo, their deity, had received the seeds of all plants from Dzhe Manido, the creator of the universe, and that Winabojo had given them to Nokomis, grandmother, the Earth, to keep in her bosom for the use of the Indians. Jack Doud also said that Winabojo took some of the native foods from his own body. He said that Winabojo pulled out a little pinch of flesh and threw it on the ground telling it to grow there as *mandamin* or corn for the Indians. Another pinch yielded squash, another beans and so on until Winabojo had very little flesh left on his body. In other words, the Indians did not know the sources of their cultivated crops, and had invented this tradition to attempt to explain their presence. (1932, 349)

This "deity," "Winabojo," of whom Huron Smith writes (called by various names in aadizookaanan, including Wenabozho and Nenabozho) is an integral part of izhitwaawin. He is a part-human, part-spirit being at the center of anishinaabe cosmology. Winabojo brought the Anishinaabeg many of the teachings that make up gikendaasowin. According to aadizookaanan, the earth as we know it today would not have existed without him. In this passage, Jack Doud alludes to one aadizookaan about how Winabojo helped to create this world. To dismiss it, casting it aside as something "invented" by the Anishinaabeg, negates a history that is at the core of gikendaasowin.

The writers of these colonized texts also deny the existence of gikendaasowin by presenting it as something that is now gone. Some of them insist on talking about izhitwaawin in the past tense, even when they are soliciting information from living Anishinaabeg who are still doing the things of which they are speaking. Melvin Gilmore consciously writes about izhitwaawin in the past tense. He explains this decision: "In noting the various uses made of plants I have employed the past tense unless I have definite knowledge of the continuance of such applications to the present" (1933, 121). Gilmore makes himself the judge of whether or not these aspects of gikendaasowin are living or dead. He gathers this information from interviews with living Anishinaabeg, who obviously carry this information with them, but he makes himself the expert on what parts of this information are still practiced. When Densmore writes of the Midewiwin, she slips in and out of the past tense. She begins one paragraph in the present tense: "It is a teaching of the Midewiwin that every tree, bush, and plant has a use." Then, in the same paragraph she slips into the past tense: "Although the Midewiwin was a repository of knowledge of herbs it did not have a pharmacopoeia accessible to every member" ([1928] 1974, 322–23). Throughout this paragraph, she continues to switch between past and present tenses when talking about the Midewiwin. Densmore acknowledges that members of this organization are living and practicing their culture at the time she conducted this research. She writes, "The men and women who at the present time (1918) treat the sick by Mide remedies are well poised and keen eyed, with a manner which indicates confidence in themselves, and which would inspire confidence in the sick persons to whom they minister" ([1928] 1974, 323). She has obviously seen living members of the Midewiwin using Mide knowledge to cure people, yet throughout her text she speaks of the Midewiwin as if it is something of the past, something that is now gone. To talk about something so important to izhitwaawin in the past tense is to say that both the Midewiwin and izhitwaawin are dead. Obviously these examples are from very old texts, the most recent of which is still over seventy years old, but these are some of the few texts we have on this subject. These texts are still used as sources of information on botanical gikendaasowin, and as such specific colonized attributes of these texts must be discussed.

## COLONIZATION AND THE COLLECTING OF GIKENDAASOWIN

Through the colonization process, non-natives from all walks of life have become "experts" on certain indigenous peoples. Throughout history, criteria for being an "expert" on a particular group of indigenous people have not been very selective. Often, any non-indigenous person having contact with an indigenous group could be considered an expert on that group. As mentioned in chapter 1, the Bureau of American Ethnology once requested that anyone having any interaction with Indian peoples send information on those people to the BAE. Many of the early reports written by these "experts" are still considered important sources of research about various indigenous peoples, just as they were when they were written. Linda Smith explains that "travelers' tales" and reports by anyone who had early contact with indigenous peoples contributed to creating an image of those indigenous peoples in the minds of other "Europeans," thus contributing to the body of research on those indigenous peoples (1999, 80–83). Officials in the colonial structure have also conducted research on indigenous peoples. Linda Smith describes the results of research by various individuals, such as military officials and land commissioners, who played a role in the subjugation and colonization of Maori people, but who were also, in later life, considered experts on those people. She writes, "Their authority as experts in Maori things was vested in the whole structure of colonialism so that while engaging in colonial operations with Maori, they also carried out investigations into Maori life that later were published under their names. Through their publication, they came to be seen by the outside world as knowledgeable, informed and relatively 'objective.' Their 'informants' were relegated to obscurity, their colonial activities seen as unproblematic, and their chronic ethnocentrism viewed as a sign of the times" (1999, 82). These researchers whom Smith describes colonized Maori knowledge while colonizing the Maori people. During this colonization, the world began to see these researchers, and not the Maori themselves, as experts on Maori knowledge.

Within the academic record of gikendaasowin, we also find examples of members of the force of colonization colonizing knowledge as well as people. Hoffman had been a surgeon with the U.S. Army during the "Indian wars" before becoming an assistant ethnologist for the BAE. He even served with General Custer (Chamberlain 1900, 44). Having had much contact with Indian people, Hoffman was hired by the BAE, the leading anthropological institute of the time, as an assistant ethnologist. As seen in the passage suggesting that "Catholic Fathers" helped to improve botanical gikendaasowin, quoted earlier in this chapter, Hoffman's text contains this "chronic ethnocentrism" of which Linda Smith writes. Although comments such as this one are often dismissed as "just a sign of the times," they are also an example of the views held by the colonizing structure of which Hoffman was a part.

Albert Reagan was also part of this colonizing structure. He conducted his research with the Anishinaabeg while working as an Indian agent at Nett Lake in Minnesota, at the beginning of the twentieth century (Reagan 1921, 246; Reagan 1922, 332; "Notes and News" 1937, 187). One of his main native consultants, George Farmer, was working as a policeman on the Nett Lake Reservation. As Indian agent, Reagan undoubtedly had some control over Farmer's employment, and Reagan admits that it took good deal of persuasion to get Farmer to translate his personal notebook of medicine songs and recipes (1922, 332). One wonders how Reagan's position as Indian agent, during a time when the U.S. Government was bombarding all Indian peoples with assimilation efforts, influenced the willingness of Farmer, and other native consultants, to work with him.

Even those not officially working for the colonizing structure were not above using coercion as one of their research techniques. Densmore, despite her reservations at recording certain kinds of information, described in chapter 1, admits to trying different approaches to get her consultants to give her the information she wants to record. She even compares working with a native consultant to "examining a witness": "An Indian may be willing to tell what is desired and not know how to express it. Sometimes one will question an Indian for a long time and the Indian will leave out the things one wants most to know; then he will suddenly give the whole information without realizing it, or in reply to a seemingly casual question. One must be like a lawyer examining a witness" (1941, 537). She adds that her consultants often resist answering questions if they feel that someone is questioning them too closely. Indian people were clearly research subjects to Densmore, as seen in this statement: "In my own work, I try to have the Indian feel that we are friends, talking over things in which we are mutually interested. In that way he becomes interested in clearing up points that I do not understand, and in the end I have the desired information" (1941, 537). This statement shows that Densmore is not above manipulating the people with whom she works in order to reach her research objectives. She will even act like a friend if she can see any benefit to her situation.

Many of the Anishinaabeg consultants who assisted researchers of gikendaasowin have been, as Linda Smith describes it, "relegated to obscurity." In "The Midē'wiwin or 'Grand Medicine Society' of the Ojibwa," Hoffman published approximately 150 pages of information on an Anishinaabe religion, but although frequently mentioning the "Midē' priests," he gives neither their names nor the names of any of his consultants. He also does not specifically state the communities in which he conducted his research, although he frequently mentions reservations in Minnesota (1891). Gilmore says that the information in his article comes from a number of Anishinaabeg living around Pinconning and Lapeer, Michigan, and near Sabrina, Ontario. He identifies the person who gave him a piece of information in only a few entries. Usually he mentions neither the native consultant from whom information came nor the location of the people who gave it to him (1933).

We do not know if these consultants, living at a time when native peoples were often persecuted for practicing their religions and cultures, requested not to be named. The authors do not tell us, but by not naming the individuals with whom they worked, researchers such as Hoffman and Gilmore are grouping all Anishinaabeg into one category. They are essentially saying that this use for a plant or this teaching is held commonly by all Anishinaabeg. Memmi refers to this practice of referring to colonized people as a group rather than as individuals as part of the "depersonalization" of the colonized. Memmi writes, "The colonized is never characterized in an individual manner; he is entitled only to drown in an anonymous collectivity" (1965, 85).

Even when consultants are mentioned, the title of "expert," whether given by academics or the general populace, generally goes to the non-native researcher. For example, many people, native and non-native, have heard of Frances Densmore, but few can name even one of the hundreds of native consultants with whom she worked across the United States, although in her texts Densmore does name, and occasionally provides pictures of, her consultants.

## COLONIZATION AND THE FRAGMENTATION OF GIKENDAASOWIN

Through the process of colonization, indigenous knowledge has been further changed by being divided and fragmented. Referring to the work of Frantz Fanon and others, Linda Smith argues that colonization disconnected indigenous peoples from their languages, their knowledge, and their ways of interacting with each other and the world. She writes, "It was a process of systematic fragmentation which can still be seen in the disciplinary carve up of the indigenous world: bones, mummies and skulls to the museums, artwork to private collectors, languages to linguistics, 'customs' to anthropologists, beliefs and behaviors to psychologists. To discover how fragmented this process was one needs only to stand in a museum, a library, a bookshop, and ask where indigenous peoples are located" (1999, 28). This is one of the dangers of taking information from one knowledge-keeping system and forcing it into the confines and categories of another. It is also a systemic effort to carve up indigenous cultures while simultaneously separating indigenous peoples from those cultures in an effort to weaken and assimilate them. Today we have indigenous peoples struggling to revive their cultures and reconnect with their languages, knowledge, and ways of interaction. Many of them turn to colonized texts for assistance. This dependence can cause many problems. First, as a result of this "fragmentation" process, colonized indigenous knowledge is not readily accessible to the people from whom it was originally taken. Even very usable material, if not easily accessible, is of little or no use to revitalization programs. The notebooks of information collected by Erminie Wheeler-Voegelin, for example, are only available to those who are willing and able to travel to Chicago, stay in the city for several days, and read the manuscripts in a research library. Although some of the published texts discussed here have been reprinted, most of them are not currently in print. Some of them have never been reprinted at all. Therefore, many of these texts are only available to those with borrowing or interlibrary loaning privileges at academic libraries. There is also the issue of time and resources to consider. One would really have to look at more than one of these texts to get enough usable material to create curriculum or exercises for a cultural revitalization program. So if one could get a copy of more readily available texts, such as the reprinted versions of Densmore texts, that person would still have to have a lot of time and resources to turn the information in those texts into something appropriate for a program revitalizing izhitwaawin.

Another result of this "fragmentation" of indigenous knowledge is that pieces of indigenous knowledge are separated into categories that follow non-native constructs rather than the constructs out of which they originally came. Anishinaabe elders gave non-native researchers information about botanical gikendaasowin. Often those researchers then divided that information into separate publications, and in some cases they acknowledge this division in their texts. For example, several times Densmore acknowledges that gikendaasowin about plants includes certain songs. In the introduction to "Uses of Plants by the Chippewa Indians," Densmore says, "The present study is related, in two of its phases, to the study of Chippewa music which preceded it. Herbs were used in the treatment of the sick and in the working of charms, and songs were sung to make the treatment and the charms effective. Songs of these classes having been recorded, the Indians were willing to bring specimens of the herbs and to explain the manner of their use" ([1928] 1974, 281). In her earlier publication, *Chippewa Music,* she says, "In the working of a charm it is considered necessary to use both the proper song and the proper medicine. For that reason a small quantity of the medicine is furnished to a person who buys such a song" (1910, 20). Even though she repeatedly describes the connection between songs and plants within izhitwaawin, Densmore divides this information into separate publications, generally not giving any indication as to what herbs are connected to what songs. For instance, in her first volume of *Chippewa Music,* Densmore writes, "The treatment of sick is conducted by older members of the Mĭdē'wĭwĭn, special songs being sung in connection with the use of medicinal herbs"

(1910, 92). Shortly after stating this connection, Densmore presents two such healing songs, mentioned in chapter 2, which were given to her by Mi'jakiya'cĭg, an elder woman who is part of the Midewiwin. These songs were used to cure Mi'jakiya'cĭg when she was a young woman and was very ill. Densmore presents the songs, translates them, and analyzes them, but she neither names nor refers to the herbs used in conjunction with these healing songs (1910, 92–94). Other researchers do this too. Albert Reagan took the information out of George Farmer's medicine notebook and divided it into two separate articles. Reagan published the songs in 1921 and the recipes in 1922. Neither publication gives an indication as to organization in the original notebook. Readers are not told whether some of these medicinal recipes and songs were to be used together.

When these authors take information from Anishinaabe people and use another worldview to divide it into categories, they render this material nearly useless to cultural revitalization programs. These are only two examples of how this information is divided in various publications. Given that these examples include information about Midewaajimowin, which from the perspective of izhitwaawin is guarded knowledge, it might be considered a good thing that this information has been presented in such a way as to make it useless to the general public. If, however, the elders who shared this information with researchers believed that they were doing so to preserve it for future generations, the presentation of this knowledge has prevented this from happening. In these publications on botanical gikendaasowin, there are examples of public knowledge having to do with inaadiziwin that is also divided into separate publications. For example, books on anishinaabe "folklore" or "myths" often contain dibaajimowinan and aadizookaanan about the origin, histories, and uses of plants,[1] but these stories are usually left out of lists of plant uses such as those discussed in this paper.

Another result of the fragmentation of knowledge is the abbreviation of gikendaasowin. Some of these authors give short descriptions of how the Anishinaabe work with these plants, providing just enough details for someone to duplicate the process. Unfortunately, the fragmented nature of these descriptions leaves out important information about how to safely handle

1. For one example of plant stories in a collection of folklore, see Laidlaw 1922, 28–38.

and administer these plants. Any medicine can be poisonous if not correctly used, and some medicines require even more caution because of the toxicity of the materials involved. A native medicine person working with these materials knows how to safely work with them. A researcher gathering a list of plants and their uses may not know or simply may not be concerned with such details. For example, Densmore provides detailed recipes for making medicines and "charms" out of plants, but some of these would not be safe to use without having some other knowledge of the plants involved. Densmore describes a cure for a cold using a "[v]ery weak decoction of roots" of *Apocynum androsaemifolium* L., commonly called spreading dogbane. She adds that this medicine is "[u]sed only for infants" ([1928] 1974, 340–41). Although someone might be able to make a medicine out of this description, it would not be prudent to do so without knowing more about this plant. Spreading dogbane is poisonous and therefore dangerous to use without proper instruction (Niering and Olmstead [1979] 2001, 346). When not handled and prepared properly, a poisonous plant can be deadly, especially to a small infant. Densmore mentions spreading dogbane several other times, saying that this plant can be used for heart palpitations and earaches, and that it can be "[c]hewed to counteract evil charms," but nowhere in the recipes for any of these medicines does Densmore mention that this plant is poisonous ([1928] 1974, 336–37; 338–39; 360–61; 376).

In his entry for *Juniperus communis* L., commonly called shrub juniper, Gilmore describes how the Anishinaabeg use this plant, saying, "The twigs and leaves were boiled to make a drink to be taken as a remedy for asthma" (1933, 124). Such a description is dangerous because someone wishing to make this medicine could very easily get some twigs and leaves of this plant, boil them, and drink them, without knowing that this plant can be deadly if one does not know how to properly administer it. In the *Field Guide to Eastern/Central Medicinal Plants and Herbs,* Steven Foster and James A. Duke give the following warning about *Juniperus communis* L.: "Potentially toxic. Large or frequent doses cause kidney failure, convulsions, and digestive irritation. In Germany, use limited to four weeks. Avoid during pregnancy. Oil may cause blistering" (2000, 254). From this description, it is obvious that there are many dangers to using *Juniperus communis* L. If someone used this plant, as suggested by Gilmore, to treat frequent attacks of asthma, that person could die from the toxicity levels in this plant. Densmore, Gilmore, and many of the other researchers discussed in chapter 1 probably did not see their texts as becoming "how-to" manuals. They also may not have known about the hazards of working with certain botanical materials, but the Anishinaabe people who gave them this knowledge most likely did. In the context of izhitwaawin, medicinal knowledge is passed down through the generations along with warnings such as who should be given a medicine, how much, and how often. Somewhere in the collection or presentation process these authors created a fragmented version of this knowledge, thereby failing to include important information about how to work with these plants.

## THE NEED FOR DECOLONIZATION

The purpose of presenting these examples of the colonization of gikendaasowin is not to slander the work of previous researchers, but to demonstrate why these texts need to be "decolonized" before they can be used in programs revitalizing izhitwaawin. These researchers conducted their research to the best of their abilities, and their publications are products of the disciplines for which they worked and the mentalities of the societies in which they lived. These researchers and their native consultants were caught up in various stages of assimilation efforts by the United States and Canada. Having seen the destruction of Indian peoples and cultures, there is little doubt that these researchers and their native consultants believed, on some level, that they were recording this information so that it could be preserved. For instance, Hoffman conducted his research at the beginning of the Dawes Era, when the Anishinaabeg, like many Indian tribes, were preparing to be moved from their reservations to small, individual parcels of land. At the end of his paper, Hoffman mentions that the native consultants with whom he is working are preparing to take allotments at the Red Lake and White Earth reservations. He says that, anticipating this move, the Anishinaabeg with whom he is working agreed to provide him with this information because they were afraid it would be lost very shortly (1891, 299–301). He explains, "The Chief Midē' priests, being aware of the momentous consequences of such a change in their habits, and foreseeing the impracticability of much longer continuing the ceremonies of so-called 'pagan rites,' became willing to impart them to me, in order that a complete description might be made and preserved for the future information of their descendants" (1891, 300). Through these researchers, our elders preserved this gikendaasowin for us. In order to use this information to revitalize izhitwaawin, however, we need to decolonize it by taking usable information out of these texts, making additions where necessary, and leaving behind degrading, ethnocentric comments made by their authors.

## DECOLONIZING GIKENDAASOWIN: BISKAABIIYANG APPROACHES

Because of the colonization process, many of us no longer see the strength of our indigenous knowledge. Our minds have been colonized along with our land, resources, and people. For us as Anishinaabeg, the decolonization of gikendaasowin is also part of the decolonization of ourselves. When describing colonization within Biskaabiiyang approaches to research, one says, *"gii-zhaanganashiiyaadizid,"* which means someone who tries to live his or her life as a non-native, at the expense of being an Anishinaabe (Anishinaabe Wordlist 2003). As stated previously, "Biskaabiiyang" means to return to ourselves, to decolonize ourselves. Biskaabiiyang research begins with the Anishinaabe researcher, who must look at the baggage that he or she carries as a result of colonization. That researcher must then rid him or herself of that baggage in order to return to inaadiziwin (Horton, pers. comm.).

Part of this decolonization is discovering or accepting that the Gete-anishinaabeg knew a great deal about the world and had the technology to survive in it. It takes great skill to be able to make a canoe out of materials found in the forest or to be able to gather and prepare wild foods. It takes perseverance and strength to gather and maintain the extensive body of pharmaceutical knowledge known to the Gete-anishinaabeg. As Anishinaabeg, we should be proud that our ancestors carried and perfected this knowledge. We should be even more proud that our elders have maintained much of this knowledge in the face of the best assimilation efforts that non-native society could throw at them. Part of decolonizing the gikendaasowin preserved in the academic record is to pick out these physical uses for plants and trees and to bring this information into programs revitalizing izhitwaawin. Unfortunately, this decolonizing is not as simple as collecting passages from various texts on how to use plants and trees and putting them together in some sort of coherent whole. These are colonized texts, and the information presented in them is generally not presented in the same way that it would be within izhitwaawin. This colonization has implications for the presentation of a plant's spiritual properties, as will be discussed later, and it also implicates the accuracy of this information. Additions and deletions must be made to this information during the decolonizing process. What follows is a brief overview of how to solve some of the problems found in colonized depictions of botanical gikendaasowin.

## SUGGESTIONS FOR DECOLONIZING BOTANICAL GIKENDAASOWIN

These suggestions on how to decolonize academic presentations of gikendaasowin include using other sources and pieces of information from the colonizers. There are several reasons for using these sources. Whenever possible, Anishinaabe elders should be our first source of information about how a certain plant or tree is viewed or worked with within the context of izhitwaawin. However, many of our elders today have not lived and worked with this information on a regular basis for a long time, and when information is not used regularly, pieces of it are often forgotten. When working with poisonous materials, it is imperative that we find out what part of what plant is safe to work with before bringing that gikendaasowin into a revitalization program, especially one involving children. There are other reasons for using these texts, however. Some of this information has been lost from our communities, and it exists now only in these colonized texts. We need to use information from our elders along with other colonized texts to help us decolonize written versions of botanical gikendaasowin. For instance, when identifying a plant, especially one our elders no longer know, we need to look at other colonized texts to find out what plant a researcher is describing, especially if she or he gives only limited identification information. Finally, our world has changed a lot since the time of the Gete-anishinaabeg, and we need to realize that things like pollution have also resulted from colonization. The colonizers caused these problems, and some of them have written a great deal about the effects of and how to work safely alongside of problems such as pollution. Resources of the colonizers, therefore, are part of the decolonization of gikendaasowin. They colonized us; their knowledge will help us decolonize our knowledge and ourselves.

## DECOLONIZATION SUGGESTIONS: ADDITIONS NEEDED

As discussed earlier in this chapter, one of the most pressing problems with these colonized texts is that they contain some inaccurate information. Some of the information in these texts has been abbreviated, possibly because these researchers spent such a short time in the field or possibly because they did not work with this knowledge on a regular basis, as one living inaadiziwin would. Abbreviated information can be dangerous, as explained earlier, because certain warnings about how to safely work with a plant or tree are generally left out of the one sentence or paragraph of information many of these researchers devote to each plant. If we are going to bring this botanical information back into the context of izhitwaawin, then part of the decolonizing process will have to include reconstructing on the basis of abbreviated information. Proper identification of plants involved is essential to reconstructing this information so that all available information about those plants can be gathered. As with the solution to most of these problems, we need to use all available resources, both non-native and anishinaabe, to do this. Keewaydinoquay often said of properly identifying plants, "*There is not room* for a mistake because dead is dead, and there are no degrees of deadness, slightly dead is just as bad as very, very dead" [emphasis in original speech] (1988). Plant identification is much easier in today's world than it was when many of these non-native researchers were conducting their research because today we have the advantage of multiple printed plant and tree identification books being available in every bookstore. We also have the information available to us on the World Wide Web, such as the database of plant identification offered by the Department of Agriculture.[2] Those using

2. The database of plant identification offered by the U.S. Department of Agriculture is available at http://plants.usda.gov.

these resources should keep in mind that just because a piece of information about the poisonous nature of a plant does not come to us from an anishinaabe source, that is no reason to disregard it. As already stated, the Gete-anishinaabeg undoubtedly knew the dangers of working with certain plants. After all, they had to use this information on a regular basis, but today some of this information has not been a living part of our culture for a long time. Where warnings about the dangers of working with certain botanical materials do not exist in colonized texts or no longer exist in dibaajimowinan, we need to add them.

In the spring of 2005, when a colleague of mine sent me a list of plants that she hoped to use with her high school students, I saw the potentially disastrous effects that texts on gikendaasowin can have when not annotated with warnings about poisonous natural materials. My colleague compiled the list of plants from a series of flash cards owned by the school where she was teaching. She did not know who had created these flash cards nor where they found the information that was on them, but they were labeled as being plants used by the Anishinaabeg. She wanted to use the information on this list with the children in her class. On one flash card was written: "Spreading Dogbane: Headache Cure." Admitting that she did not know about the plants on this list, she sent me this list in hopes that I might be able to give her more information. I alerted her to the dangers of using spreading dogbane, described previously. If she had not thought to learn more about the plants on this list before sharing this information with her students, the results could have been deadly. There is no way of knowing what telling a class of students about this plant without telling them that it is poisonous could do. Some students might try out a cure learned in such a class, unintentionally harming themselves or others. It is imperative, therefore, that anyone wishing to use the information from a book on American Indian medicine should learn as much as possible about the plants listed in the text before attempting to use them. If these written texts are going to be used in cultural programs, then resources are needed that present information about gikendaasowin, but that also provide warnings about the gathering and preparations of certain hazardous materials.

We should also remember to find as much information about the levels of pollutants in various plants as possible. Those who think that the knowledge of pollution and its effects on our environment are not a part of izhitwaawin should know that when speaking from the perspective of izhitwaawin, Keewaydinoquay, Dora Dorothy Whipple, and Mary Geniusz all warn about the dangers of pollution. In one example, Keewaydinoquay warns that heavy metals have altered the growth pattern of common mullein (*Verbascum thapsus* L.). Mullein used to have only one flowering stalk, or head, but altered mullein often have multiple heads. Keewaydinoquay says that one should not use these altered mullein for any purpose because we do not know what else has changed within the mullein along with its appearance (1990a). Whipple warns that people should not use giizhikaandag (cedar) that is growing in the city. She says that pollution has affected these trees, and that one should only use giizhikaandag growing in the country (pers. comm.). Mary Geniusz often speaks of the many virtues of cattails, including how they can be used as insulation and eaten as food. She warns, however, that cattails like to soak up anything that is around them, and they are particularly good at internalizing pollutants. One should never pick and use cattails growing along a roadside or near some other source of pollution, such as a factory (pers. comm.). Warnings about pollutants have become part of izhitwaawin. This fact should not surprise us, as izhitwaawin is not static; it does change. All living cultures do change. Dangers about pollutants, however, were nearly unknown when many of these researchers wrote their texts on gikendaasowin. These warnings must be added to a decolonized text.

## DECOLONIZATION SUGGESTIONS: DELETIONS NEEDED

The decolonizing process should also discard or ignore certain portions of these colonized texts. As exemplified earlier, some of these texts include degrading comments not only about gikendaasowin but also about the Anishinaabeg. References to the Anishinaabeg being primitive, simplistic, or somehow lacking intelligence need to be left to rot in the colonized pages on which they were written. Working with language and culture revitalization is a difficult enough process. Those working in these areas do not need further discouragement in the form of degrading comments made by researchers who have been gone for decades. One of my interests in decolonizing this literature is so that other Anishinaabeg, especially members of our younger generations searching for pieces of our culture, will not stumble upon these insults because some colonized version of izhitwaawin was all that was available to them. Whatever the authors' original reasons for including these insulting passages, they have no place in our revitalization efforts. We can use information from these texts, such as information on working with botanical materials or stories explaining the origin of certain plants, but we need to treat this knowledge with the proper respect by disentangling it from belittling statements made by the researchers who collected it. These comments have fueled systemic racism and oppression long enough.

As explained earlier, some of these sources further degrade Anishinaabeg by speaking of gikendaasowin and izhitwaawin as if they have completely disappeared. Gilmore and Densmore, the two researchers cited earlier who use the past tense, were both writing at a time when many Anishinaabeg were living in great poverty, surrounded by a society that had already tried multiple ways to assimilate them. Gilmore and Densmore might have seen their work as archiving and preserving a dying culture. By the time Densmore published "Uses of Plants by the Chippewa Indians," twelve of the Anishinaabeg who gave her the information presented in this text had died ([1928] 1974, 282–83). Gilmore says that he just did not see the Anishinaabeg using certain medicines anymore (1933, 121). Researchers like Gilmore and Densmore had their own reasons for employing the past tense in their articles, but for the purposes of revitalizing izhitwaawin, speaking of using plants and trees as if we no longer do these things simply will not work. At the beginning of the twenty-first century, these presentations of the death of izhitwaawin are discouraging and insulting to those of us who now know that anishinaabe culture is not dying. This way of talking about anishinaabe culture seeps into the speech of Anishinaabeg themselves—another result of colonization. We began to see ourselves as existing only in the past. When I hear Anishinaabeg speak of izhitwaawin in the past tense, I wonder how much of this tendency to speak of our traditions as if they were gone comes from the work of non-native researchers, who insisted upon writing about our culture as if it were dead and buried. These depictions of the Anishinaabeg and izhitwaawin must be separated from the valuable information in these texts if they are to be used for cultural revitalization.

## DECOLONIZATION SUGGESTIONS: CHOOSING SOURCES OF INFORMATION

As a resource for the revitalization of izhitwaawin, texts that do not document the sources of the information they hold are harder to recontextualize in accordance with izhitwaawin. Different anishinaabe communities hold different teachings, and it is important when revitalizing any gikendaasowin within a community to know whether or not the teachings being read about in a text are the same as those held by that community. Of course, it is always advisable to consult with elders in a community to see if the information about to be taught in a cultural revitalization program supports or conflicts with their teachings, but when dealing with gikendaasowin that has been lost, the process is much more difficult.

This difficulty is not simply an issue of not knowing to which elder goes the proper credit for these teachings. It is also an issue of determining which pieces of gikendaasowin actually came from an Anishinaabe. We should continue to be aware of encountering sources such as Gerald Stowe's article, "Plants Used by the Chippewa," which claim that the information presented in them comes directly from the Anishinaabeg, but names neither a person nor a community from which that researcher gathered this information. As explained earlier, Stowe's article appears to be a copy of Densmore's text, which did come from Anishinaabeg, but an article such as this with no sources mentioned could just as easily have been a fabrication of the author. In addition, as mentioned previously, the authors of *Plants Used by the Great Lakes Ojibwa* do not specifically state the sources for the uses of plants found in that text. The fact that they often describe those using these plants as "the Indians" rather than "the Ojibwa" suggests that this information could describe how any group of Indians on the continent uses these plants. With the exception of these two texts, all of the authors mentioned in this research do appear to have actually spoken to Anishinaabeg and collected their information from them. Clues that these texts are not complete fabrications of their authors' imaginations are: many of them mention anishinaabe communities where they conducted their research; some of them mention the names of specific native consultants and give exact years in which their research was conducted. They also contain similar information.

This similarity is not always the case, however, and we should be aware of it. There exists a rather large body of literature claiming to be collections of "legends," "tales," or other stories that come out of izhitwaawin. Those conducting ethnological research on the Anishinaabeg wrote some of these. In some cases, a researcher would publish one ethnographic text on the Anishinaabeg and another on their "folklore." Paul Radin, for example, published *Some Myths and Tales of the Ojibwa of Southeastern Ontario* and later published "Ethnological Notes on the Ojibwa of Southeastern Ontario" (1914; 1928). In other cases, the author of a book of "legends" claims to have heard these stories from the Anishinaabeg, but names neither an individual nor a community. Such collections must be suspect. Nearly all collections of anishinaabe "legends" are written in English, occasionally with a few scattered Ojibwe words. Most of these are also only a paragraph or two in length. Plants and trees do come up quite a bit in these stories, and they could be revisited and made part of the revitalization of izhitwaawin. Those conducting this research, however, should be aware that anishinaabe stories have long been of interest to those writing children's literature. Children's authors have rewritten these stories to suit younger audiences, but they seldom acknowledge where they first heard or learned of this story. Because these are not scholarly works, they are not part of this research, but those interested in retelling these stories for cultural revitalization programs should be cautious when using them as many of these stories have been greatly altered from the stories originally collected from our elders.

## THE DECOLONIZATION OF OJIBWE PLANT NAMES

Language is another important component of the decolonizing process. Anishinaabemowin and izhitwaawin are closely connected. Some elders say that Anishinaabemowin is the vehicle of izhitwaawin and without it we will lose large portions of izhitwaawin. For example, some dibaajimowinan say that certain ceremonies have to be conducted in Anishinaabemowin. Programs revitalizing izhitwaawin often have language components. Although the information in these colonized texts is primarily presented in English, there is some Anishinaabemowin found throughout them, in the form of songs, prayers, and names of plants and trees. Given that Ojibwe plant names are one area of Anishinaabemowin in which certain pieces of information have been forgotten, these lists of Ojibwe plant names are often very important to programs revitalizing izhitwaawin. Some work would have to be done, however, to decolonize these plant names so that they could be usable to students and teachers of Anishinaabemowin.

Lists of Ojibwe plant names found in these texts have varying degrees of usability to anishinaabe cultural revitalization programs. For instance, the writing system Densmore uses to record the Ojibwe names in her text makes them difficult to decipher properly. When describing her writing system, Densmore acknowledges that she hears certain sounds in the Ojibwe language that she chooses not to record when writing Ojibwe words ([1928] 1974, 283). All of the sounds that Densmore chooses not to indicate in her recordings of these words, including nasalization, vowel length, and "an obscure sound resembling *h* in the English alphabet," are crucial to the pronunciation of these words. Fortunately, Densmore does provide scientific names, and from these scientific names it is at least possible to identify the plants in Densmore's text; and with these identifications, it is easier to figure out Densmore's recorded Ojibwe names by comparing them to those found in other written and oral sources. She also provides translations of many of the plant names given in her text, making it possible to reconstruct some of these words based on these English translations.

In part 2 of John Tanner's captivity narrative, Edwin James also lists Ojibwe plant names, but without more information many of the plants in this list are nearly unidentifiable. For example, in one entry the author writes, "Muk-koose-e-mee-nun—Young bear's berries" ([1830] 1956, 298). Here, "Young bear's berries" could be a common English name, but it could also be a literal translation of the Ojibwe name. "Muk-koose" sounds like it comes from *makoons*, which literally means young bear, and "Mee-nun" sounds like *-minan*, a common final found on Ojibwe plant names, one literal translation of which is "berries." Whether "young bear's berries" is also a common English name for this plant is unknown. Although it appears to be possible to reconstruct this Ojibwe name, we still cannot identify this plant, as Ojibwe and English names vary with location, and we do not know the exact location from which this name comes. Also, we have no other information about this plant and almost all of the others in this list, such as what it looks like, where it grows, or how the Anishinaabeg use it.

Huron Smith includes Ojibwe names for many of the plants in his text. He claims to have two books written in the Ojibwe language, including a dictionary written by Bishop Baraga. Smith's pronunciation key reflects that he understands there are long, short, and nasalized vowels in the Ojibwe language. Whether he can distinguish between these sounds when hearing and recording Ojibwe words, however, is unknown. The accuracy of Smith's transcriptions of Ojibwe words must be questioned, especially considering the following statement in his foreword: "While the writer is not a linguist, Indian pronunciation came easily to him and he was able to pronounce all plant names in an intelligible manner to Ojibwe people whom he had never seen before" (1932, 336). Anyone familiar with studying languages should see the problem with Smith's statement. There is a tremendous difference between being able to say a word "in an intelligible manner" and being able to say, or write, that word accurately.

Gilmore's article also includes some Ojibwe plant names. Although he provides a pronunciation key, he gives no indication of the degree of his Ojibwe language proficiency, and so there really is no way of knowing if he was capable of accurately transcribing the names given to him. Gilmore's pronunciation key indicates that he is aware of some of the important distinctions of sounds in Ojibwe, such as long and short vowels, nasalization, and various consonant clusters, but once again, we cannot judge his ability to distinguish between these sounds (1933, 122).

In *Plants Used by the Great Lakes Ojibwa*, Meeker, Elias, and Heim include Ojibwe names in their text, but the names they include come from a variety of researchers, including some of those just described. Although some of the Ojibwe names in this text have already been checked with Ojibwe speakers and retranscribed, the majority of them appear in this text just as the original researchers who gathered them wrote them. Therefore, these names retain the questions of accuracy and pronunciation given in the descriptions of names recorded by Densmore, Gilmore, and Huron Smith. The authors of *Plants Used by the Great Lakes Ojibwa* do not include in their text pronunciation keys for any of these other researchers' writing systems (Meeker, Elias, and Heim 1993a). In the abridged version of this text, they do not provide important information that would be helpful to anyone attempting to figure out these words, namely the accent marks found in the original texts and the standardized names found in the body of the unabridged version (1993b). In the original texts, some of the authors provide English translations for the Ojibwe names they list, but none of these translations appears in either version of *Plants Used by the Great Lakes Ojibwa* (Meeker, Elias, and Heim 1993a; 1993b). Anyone wishing to find these pronunciation keys or translations would have to look in the articles and books from which these names were gathered, most of which include some sort of pronunciation key. These original sources are often obscure and difficult for anyone to reach unless he or she has access to an academic library.

Some of the Ojibwe names for plants and trees found in these colonized texts appear to be recoverable; others do not for the reasons just described. The best way to go over these names is to work with an Ojibwe speaker to check them for accuracy. I have done this check for all of the plants listed in *Plants Used by the Great Lakes Ojibwa,* and, although Rose, the speaker with whom I was working, did not know all of the names for every plant listed in that text, she did know quite a few of them. In some cases she had names different from those listed; in other cases she had the same name and was able to help me make corrections to the names for proper vowel length and pronunciation (Rose, pers. comm.). For people interested in going over colonized versions of Ojibwe plant and tree names, it is important to understand that the same plant can have several different names in Ojibwe. George McGeshick, Sr., says that Anishinaabeg from different communities have different names for the same plant, and often these names are based on how those people use that plant (pers. comm.). Because names for plants and trees do vary in different communities, getting local names for plants from the elders connected to a program revitalizing izhitwaawin is important. Names found in colonized texts can serve as a memory-sparking device for elders who have trouble remembering a certain plant's Ojibwe name. They can also serve as a tool to reconstruct a name that has been lost. If the original recorder of a name provides a translation for that name, then even a poorly transcribed plant name can often be sounded out and reconstructed with the help of an Ojibwe speaker. Often Ojibwe plant names are compounds, containing two or more Ojibwe words with meaningful suffixes or prefixes.

When an analysis of component parts is made of these words, then they can often be reconstructed. Take, for example, the following entry that Zichmanis and Hodgins include in their list of Ojibwe plant names: "Mukikeebug." They translate the parts of this word as "frog petal" (1982, 263). When analyzing this word, it is clear that this name is a compound of *omakakii* (frog) and *-bag,* which means leaf or petal. Given this information, when written in a standard orthography, this word becomes *omakakiibag.*

## SUGGESTIONS ON WHAT TO INCLUDE IN A DECOLONIZED TEXT

It is my hope that, as Anishinaabeg become increasingly aware of the problems of these colonized texts, those already working with these texts will take steps to decolonize them. A number of questions about the decolonization process arise, such as: How much information should be included in a decolonized presentation of gikendaasowin? Should we go through every source ever written on gikendaasowin and pull out every shred of information that we can find on plants and trees? When working with a specific community from which a substantial amount of material has been collected, one might want only to decolonize materials that were originally appropriated from that community. Those trying to more generally decolonize botanical gikendaasowin might want to take a piece of advice from Keewaydinoquay, who did not teach every single use for every single plant. She said that keepers of anishinaabe plant knowledge, such as members of the Midewiwin, taught their oshkaabewisag the chief virtue or virtues of each plant, and that is how Keewaydinoquay structured her teachings. In the recordings of her classes, she often criticizes herbals as being just one big long list of every possible thing one could do with a given plant. She says that some plants can do many things, but, especially for endangered plants, one should reserve them for what they do best. For example, in one class she reads from a herbal the many uses listed for butterfly weed (*Asclepias tuberosa* L.). After reading this list Keewaydinoquay says,

> Well, see that's just a running through for everything, and there's not *enough* of this in anyplace *I've ever seen* in the Midwest in order to use it for all of those, but it is true that this particular species, *Asclepias tuberosa,* also called

> butterfly weed, is particularly good for pulmonary troubles and especially fine for pleurisy, which is an infection in the lining of the lungs . . . and it usually manifests itself in the lower part of the lungs in the back, and some people who have continual back ache in that area really have . . . chronic pleurisy. And, because I am who I am . . . I need to say that I would think that this would be the only excuse for pulling up this herb and using it, unless for some reason or other you had raised a whole lot of it, in your own yard or something or other like this. (1985) [emphasis in original speech]

Not every plant is as sparsely populated as butterfly weed, and many plants do have several things at which they excel. Keewaydinoquay often mentioned more than one use for the same plant, especially if it was a common plant found over a large area. Examples of her many teachings on cedar and birch can be found chapter 4. Still, each plant is not necessarily good at all the things it can do. Sometimes another plant works even better, and is just as or more readily available.

Many of the non-native researchers collecting gikendaasowin compiled laundry lists of the uses of plants, possibly modeled off the non-native herbal tradition, while others only recorded one or two obscure uses for certain plants. Both of these recording methods are examples of the biggest differences between anishinaabe and non-native knowledge-keeping systems. Within izhitwaawin, plants and trees are beings with spirits; in other traditions, they are objects to be studied. For example, consider the following entry on goldthread (*Coptis trifolia* [L.] Salib,) written by Gilmore: "The roots were used to make a yellow dye" (1933, 130). Gilmore provides no more information. From this passage, it is clear that Gilmore has not tried this recipe, and it is probable from similar passages throughout his article that he has not tried any of these recipes. Goldthread does indeed make a nice yellow dye, but so do many other plants that are much easier to gather. The roots of goldthread are very small, and an awful lot of them are needed to make a dye. In their field guide, Foster and Duke describe the minuscule size of goldthread roots saying, "Today it is seldom used and rarely present in the herb trade, probably because the root is literally a thread, hence difficult to harvest in quantity" (2000, 42–43). Other materials that make just as fine a yellow dye, including the flowers of goldenrod (*Solidago canadensis*

L.) and the pith of sumac (*Rhus hirta* [L.] *Sudworth*), are much more abundant and take much less time to gather and prepare. Goldthread has an even more important virtue to share with the Anishinaabeg than just being used as a dye: goldthread can take away the pain of a toothache and cure an infection in the mouth. To use goldthread this way, one makes an offering to this plant, explaining for whom these healing abilities are needed. Then, after picking the plant and cleaning the roots, one puts a piece of goldthread roots alongside the infection or toothache. In a matter of hours, the pain will be gone and the infection will start to be cured (M. Geniusz, pers. comm.; Keewaydinoquay 1985). Although there are many plants that can make a yellow dye, there are not so many plants that can treat toothaches and infections in the mouth. Someone living with this plant and relying on it to cure the pain of a toothache would not waste it by uprooting it to make a dye because such an action might render the plant too scarce to gather when the need arose to use it as a medicine.

When I began this project, I envisioned collecting little pieces of information about each plant from a variety of sources. Keewaydinoquay and my mother taught me that within izhitwaawin plants and trees have spiritual properties, songs, stories, and talents to share with us. So I was not just looking for botanical information on these plants, I was looking for everything I could find—stories, songs, recipes, Ojibwe names, everything. It was an ambitious project, and it produced a mountain of notes and citations. When I began also compiling Ojibwe names for certain parts of plants, Rose finally told me that I was thinking like a non-native. She says that in Ojibwe we do not have to have a name for every single little part of a tree or a plant (Rose, pers. comm.). English classification categories are not always transferable to other languages, and not every word in English translates into other languages. So when working with decolonization, we need to make sure that we are not simply creating new colonized materials. Encyclopedic resources listing every song, every story, and every use of every plant included in gikendaasowin may be interesting to read, but they will not really help us with the revitalization of izhitwaawin. As soon as those working in these programs have to start sorting through page after page of material in order to make lesson plans or to pick topics for their programs, they will most likely be overwhelmed with this huge amount of material.

When choosing what information to include in a revitalization program, the elders who have contact with that program should be consulted about what plants or trees they use and are willing to discuss. This information should be a major component of that program. A second choice of information could be those plants with long detailed descriptions found in the colonized record. If an author just lists a use for a plant but gives no specific instructions, it would be hard to reconstruct that piece of information. If, however, an author lists a recipe for working with a certain plant, or includes specific dibaajimowin received from native consultants about that plant, then that knowledge would be much easier to reconstruct. The presence of more detailed information on a plant or tree could indicate that there simply was more knowledge to gather and that plant or tree is more widely used in izhitwaawin.

More "modern" information should also be included in a revitalization program or a decolonized text. Decolonization does not necessarily mean a complete dismissal of non-native or "modern" information and technology. As stated previously, living cultures do change. To deny changes made to izhitwaawin is to deny the continuation of our culture. As Linda Smith argues, the refusal to accept the change of indigenous culture or to try to determine what aspects of those cultures are "authentic" is yet another product of colonization (1999, 74). According to Keewaydinoquay, Nodjimahkwe often modernized recipes. She would look at the old recipes, decide what was available to her at that time, and experiment with the recipe until she found something else that was easier to get but that would work the same. When Nodjimahkwe was younger, Keewaydinoquay explained to Mary Geniusz, she was more mobile and her people had a larger area from which they could gather plants, but as she got older both her mobility and her people's territory decreased. To continue making medicines, she modernized (Geniusz, pers. comm.). Keewaydinoquay started working with Nodjimahkwe in the late 1920s, and at that time Nodjimahkwe was already modernizing recipes (Keewaydinoquay, n.d.*c*). So this process of adapting new technologies to gikendaasowin is not a recent development. Nodjimahkwe was doing it over seventy years ago, and we know from the extensive record on the fur trade that as soon as iron knives, pots, and axes were available, Indian people began using these new technologies. Today our elders use modern conveniences such as blenders, supermarket bags, and pocketknives to gather botanical materials and make medicines. Some of them even use ingredients from the supermarket, such as seen in the recipe for "Cedar Lemon Balm" described chapter 4. We only continue to colonize ourselves when we insist on looking for the "traditional" way of working with plants and trees, while ignoring the gikendaasowin that our elders have to share with us just because it involves the use of technology unknown to our ancestors.

## CONCLUSION

Biskaabiiyang methodologies require that researchers begin their research by analyzing the teachings and philosophies within themselves, separating those of the colonizer from those of inaadiziwin. When "returning to ourselves," we must center ourselves in the dibaajimowin and aadizookaan that we receive from our elders, and from that stance we can decolonize ourselves and our knowledge, and then conduct meaningful research. When looking at my own background through Biskaabiiyang research methodologies, I thought of the dibaajimowin and aadizookaan about two trees central to izhitwaawin that my mother and Keewaydinoquay told me when I was a child. These trees are *giizhikaatig*, also called northern white cedar (*Thuja occidentalis* L.), and *wiigwaasi-mitig, also called birch (Betula papyrifera* Marsh.). I found that deep inside me, I did agree with the academic belief that the stories, songs, and teachings I knew about these trees were just not advanced or authentic enough to be a major part of this research. The colonized portion of my brain assured me that they were "fairy tales" meant to entertain children. I learned these stories and teachings in English, and my academic training taught me to question the accuracy of native stories told in a European language and following non-native storytelling traditions. When I recognized my prejudice against these dibaajimowin and aadizookaan as psychological trauma of colonization, I was able to accept these teachings and engage their help to decolonize this portion of gikendaasowin. The next chapter presents this information. The Biskaabiiyang approach used to decolonize this research will be discussed throughout the chapter in hopes that others will be able to use a similar model to decolonize research done on their own people.

# 4

# Giizhikaatig miinawaa Wiigwaasi-mitig

## *A Sample of Decolonized Anishinaabe-gikendaasowin*

### INTRODUCTION

When teaching within the context of izhitwaawin, one makes sure that those learning how to work with botanical materials can properly identify the plants and trees from which those materials come. As explained in chapter 3, there are poisonous plants and trees that injure or kill a person if used incorrectly; so, learning how to recognize plants and trees is important. It is impossible for me to take all of my readers out to learn how to identify the two trees discussed in this chapter, giizhikaatig (white cedar) and wiigwaasi-mitig (paper birch), so pictures of them, drawn by Annmarie Geniusz, are presented on the following pages. A third plant, makwa-miskomin (bear-berry, *Arctostaphylos uva-ursi* [L.] Spreng), is also pictured here because it is mentioned several times in this chapter and in chapter 2.

This chapter is a sample of the decolonization process. I begin with a presentation of the gikendaasowin I carry from my background so that the reader can better understand my focus in this research. The first piece of gikendaasowin that I ever received about the cedar tree was a song, which Keewaydinoquay taught me when I was approximately six years old. She says this song is "commonly sung when the people bless themselves. Cause . . . everything we have, every good medicine we have, whether it's spiritual good medicine or whether it's good medicine for a physical ill . . . contains something of the grandmother cedar" (Keewaydinoquay n.d.*b*). This song is

2. White cedar, giizhikaatig *(Thuja occidentalis).*
© Copyright Annmarie Geniusz. Used with permission.

an aadizookaan, that is, it has a spirit that knows that it is being sung. Aadizookaan are important keepers of gikendaasowin, some of which can be seen in the Ojibwe and English lines of this song. Here is a written transcription of this song, which Keewaydinoquay often called "Nookomis Giizhik: The Cedar Song," with both the English and Ojibwe verses, presented as she taught them to me:

giwaabamaa giizhik
mitig azhitwaa

3. Paper birch, wiigwaasi-mitig *(Betula papyrifera).*


onesenodaawaan, bimaadiziwin
gaa-noojimowaad, anishinaabeg
nindinaa nookomis, nindabandendam

*Refrain:*
nookomis sa giizhik, gichitwaawendaagoziwin

behold the cedar
the holy tree
come let her breathe you, new life within
we call her saving tree, she saves the people
we say nookomis, we show respect

4. Bearberry, makwa-miskomin (*Arctostaphylos uva-ursi* L.).
© Copyright Annmarie Geniusz. Used with permission.

*Refrain:*
nookomis sa giizhik, gichitwaawendaagoziwin

When conducting this research, I went through the Ojibwe verses of this song, and found that the English verses were not an exact translation of the Ojibwe ones. For instance, some of the verbs in the Ojibwe verses are conjugated for different subjects than are implied in the English verses.

One can, however, get a general idea of what the Ojibwe verses mean from the English verses.

The next piece of gikendaasowin that I used to help me in this research is a dibaajimowin from Keewaydinoquay, which I heard many times as a child. Geniusz says that Keewaydinoquay often referred to this dibaajimowin as "Traditional Anishinaabe Advice to Youth" or "Ojibwe Kindergarten" (M. Geniusz 2005, 46; M. Geniusz, pers. comm.). The version presented here comes directly from a recording of Keewaydinoquay addressing a university class:

> At one time when our elders did all the instruction of the young, they would tell the children that did the necessitudes of life bring them to a situation where they might be lost, or washed ashore from a raft, or carried in a canoe by a storm to a land they didn't know, put in a situation where they had no idea what their location was, what direction home was . . . they said not to worry. Go to the nearest high spot and look around and if you can see grandmother cedar and grandfather birch, you're with your family and you're safe because although you may not have the finest, most exciting time of your life, the easiest plushest time of your life, they can provide for you everything that you need for survival. Because those together will provide food, and shelter, and material to construct what you need, and water, and anything else that you need, medicines, antiseptic, anything that you need. (Keewaydinoquay 1988)

Keewaydinoquay told Mary Geniusz that once children learned this dibaajimowin, they were taught how to work with these two trees to make everything that they would need in a survival situation (M. Geniusz, pers. comm.).

The next two pieces of gikendaasowin that I know about these trees are two aadizookaanan, presented on the following pages. These are versions that I heard as a child, but that is appropriate because the transmission of gikendaasowin within the context of izhitwaawin begins in childhood. When relearning or learning these things for the first time, we must put ourselves in the place of a child. These are not "fairy tales," but aadizookaan, stories that have living spirits who know when these stories are being told and who, when respectfully asked, will help teach both the storyteller and his or her audience to learn from these stories.

I realize that I may receive criticism from my fellow Anishinaabeg and academics for presenting these stories in my research, and before proceeding, I want to address some issues. First, I do not claim that these aadizookaanan exist in all anishinaabe communities. These aadizookaanan may be similar to or far different from those told in other communities, but they are the ones I know. Biskaabiiyang research methodology requires that I, the researcher, begin with my own teachings and use them to guide my research. That is what I am doing.

Second, there are elders who say that aadizookaan can only be told in the winter and should not be written or recorded. Geniusz says that Keewaydinoquay gave her this teaching, but she also told her that stories have always been told when the gikendaasowin within them is needed. Keewaydinoquay told these stories at various times of the year in her university courses and whenever she was teaching about these trees (M. Geniusz, pers. comm.). In response to the second point, I was not the first person to write or record these stories. Keewaydinoquay made an audio recording of the first of these stories, "The Creation of Nookomis Giizhik," and she had one of her oshkaabewisag tell it on the video *Native American Philosophy and Relationships to Plant Life*. Densmore provides a written version of the second story, entitled here "Nenabozho and the Animikiig" ([1928] 1974, 381–84). Mary Geniusz provides written versions of both of these stories in her text (2005). I should add that I looked at all of these recorded versions of these stories before retelling them here, rather than relying solely on my memory of information I was given as a child.

Third, in response to those academics and Anishinaabeg who question the accuracy of native stories told in a non-native language and according to non-native storytelling traditions, I present these aadizookaanan in English to assist Anishinaabeg who do not yet speak our language in the decolonization process. When I needed to learn these aadizookaanan, they were given to me in English because I did not speak much Ojibwe. Learning Anishinaabemowin is an important step in decolonization, but learning the gikendaasowin contained within these aadizookaanan might give those attempting to decolonize themselves the necessary incentive to continue that process, as they did for me.

Finally, for those who think that stories should not be part of "scholarly research," the spirits within these aadizookaanan deserve more respect than to have their stories delegated to an appendix or a mere summary. These aadizookaanan bring to the reader concepts that form the foundation of gikendaasowin. Rather than an interruption to this academic text, these aadizookaanan introduce readers to concepts and teachings without which one cannot hope to understand the decolonization of gikendaasowin.

Retellings of these two aadizookaanan, "The Creation of Nookomis Giizhik" and "Nenabozho and the Animikiig," are presented on the following pages. A few references to these stories are made in chapter 2, and more references will be made in this chapter.

## THE CREATION OF NOOKOMIS GIIZHIK

*Gichi-mewinzha gii-oshki-akiiwan,* long ago when the earth was young, all the beings of Creation knew who they were and what was expected of them.[1] There were those who lived in the outer air, including the bineshiinyag (the birds) and the winged ones who fly around the earth. There were those who lived in the waters, such as the *giigoonyag,* the fish and the *mishiikenyag,* the turtles. There were those who lived beneath the earth, those who lived in the ground, and those who burrowed beneath the earth for periods of time. There were also those who lived on the earth. All the beings of the Creation knew that they were supposed to take care of one another, for all parts of Creation are interconnected and dependent on one another to keep the balance of the earth.

There came a time, however, when one group of beings who live on the earth, the Anishinaabeg, were in trouble. We do not know what exactly this trouble was, but it was so great that there was a chance that the Anishinaabeg were not going to survive much longer.

The creatures of the outer air saw what was happening to the on-the-earth creatures and they saw how the Anishinaabeg, the last of these beings to be created, were suffering. They saw that the Anishinaabeg, who had

1. This is a retelling of an aadizookaan that I heard as a child from Keewaydinoquay and her oshkaabewisag. The version presented here is based on two recordings that Keewaydinoquay made of this story. In one of these versions she tells the story herself (Keewaydinoquay n.d.*b*). In the other she supervises the telling (Keewaydinoquay 1986).

neither feathers nor wings nor claws, were not prepared for life on the earth. The Anishinaabeg were cold and hungry, and had trouble moving from one place to another. They seemed to walk around among the trees not really sure of what they were supposed to be doing. Knowing that Gichi-manidoo would not be happy if they let the Anishinaabeg perish, the creatures of the outer air decided to get the rest of the beings of the earth to work with them to help. So they began trying to tell the under-the-water beings and the under-the-earth beings what was happening, but they were not successful. Every time *Migizi,* Eagle, would fly low and skim across the water trying to warn the giigoonyag of the problem, the giigoonyag would dive deep into the lakes and streams. Every time *Gookooko'oo,* Owl, would swoop down in the evening to tell the *waawaabiganoojiinyag,* the mice, the problem, the waawaabiganoojiinyag would scurry under the nearest clump of grass and hide. And every time *Gekek,* Hawk, would come down from his perch above the field to try and tell the *waaboozoog,* the rabbits, of the problem, the waaboozoog would jump into their holes.

So the beings of the outer air decided they needed a messenger who could talk to the beings who live in the water and the beings who live under the earth. They chose *Amik,* Beaver, who spends his time on the earth, in the water, and burrowing under the earth, but try as they might, they could not get Amik to listen to them. The bineshiinyag sat around his pond and sang to him. They flew over Amik and dropped nuts on his head. Nothing worked. Amik just kept on chewing down logs and building his *amikwiish,* beaver lodge. They were just about to give up when *Ogiishkimanisii,* Kingfisher, said that he would get Amik to listen to them, and Ogiishkimanisii swooped down over Amik and grabbed the fur on the top of his head with his talons. The place where Ogiishkimanisii grabbed his fur is still visible on all *amikwag,* beavers. When the amikwag dive under the water and come up again, a part can be seen down the middle of their heads, right where Ogiishkimanisii grabbed Amik's fir. Amik still did not listen to the beings of the outer air. He had to finish his amikwiish, and he wanted to enlarge his pond. There were just not enough hours in the day for him to complete his work. He had no time to worry about anything else. So, Ogiishkimanisii took another dive at Amik, but this time Amik was ready and dove under the water.

When Amik stuck his head out from under the water, Ogiishkimanisii was waiting for him, prepared to make as many dives as necessary to get Amik to listen. Seeing this, Amik finally asked what it was that Ogiishkimanisii and all the creatures of the upper air wanted. They told Amik about the problems the Anishinaabeg were facing.

Amik listened to the winged ones and agreed with them that the Anishinaabeg were in a pathetic state. He voiced his own opinion on their state, saying, "They don't even have nice flat tails to use to warn each other about danger!" The winged ones, knowing how proud Amik is of his tail, nodded their agreement to his statement. Some of them casually flashed their tails at him, showing him that they too had nice flat tails.

"Well, I can see that the Anishinaabeg have trouble, but I don't understand what I can do about it," Amik finally said.

"We need you to be a messenger for us," explained the winged ones. "We need you to tell the beings who live in the waters and beneath the earth that the Anishinaabeg need our assistance."

"Now why would I want to do that?" asked Amik. "I have seen these Anishinaabeg myself from time to time, and I really don't see much use in helping them. They are kind of clueless and really don't have any impact on my life at all. Why, not one of them has ever offered to help me move a log to my amikwiish, and I have never heard any of them helping any of the other animals either. Why should we help them?"

"Remember," said the winged ones, "when Gichi-manidoo made us all promise that we would help each other to maintain the balance of this world? If we let the Anishinaabeg destroy themselves, then we will be breaking our promise to Gichi-manidoo."

Amik was about to say something about how old-fashioned that way of thinking was, when Ogiishkimanisii stepped closer to him with a stern look on his face. "All right!" said Amik. "I'll give your message to the beings who live under the earth and in the waters." He set off to contact the rest of Creation, making a big show of what he was doing. He exclaimed every few feet, "Well, I'm off to find the beings who live under the earth and in the waters . . . " He wanted Ogiishkimanisii to know that he was going to complete his task, just in case Ogiishkimanisii was thinking of taking another dive at his head.

Amik found the beings who live under the earth, and he found the beings who live in the waters, and he told them about the problems that the Anishinaabeg were having, and it was decided that all of Creation should have a council meeting to decide their course of action. First of all, they had to decide where this meeting should take place. They discussed many possible meeting places. Someone suggested the place of eternal fire, but some of the other animals said, "No, I am afraid of that place. It is very hot, and there is a weird smell there. I am afraid that if I fall asleep there, I may never wake up." Many places were discussed, until they decided to have their meeting in the place of winter slumbers.

The council meeting was very long because, like meetings today in *anishinaabewaki,* anishinaabe country, everyone had a right to be heard. Some of the animals gave very long speeches about why it was not their problem to worry about the Anishinaabeg. Other animals in equally as long speeches said why it was important to keep their promise to Gichi-manidoo to help the rest of Creation and thereby keep the balance of this world.

Finally, just when everyone was getting very hungry and tired, one of the animals had a suggestion. "Why don't we appoint a committee?" The other animals eagerly agreed. In fact, this was the first thing they all unanimously agreed on during the entire meeting. Now came the question of who should be on this new committee. Among the animals who had given the longest speeches on why they should bother to help the Anishinaabeg were *Mishi-makwa,* Bear, and *Nigig,* Otter. So the other animals said to Mishi-makwa and Nigig, "Well you two keep coming up with all these reasons why we should help the Anishinaabeg. So we are going to make you two the committee." The meeting was adjourned.

Now this was at a time when not too many *makwag,* bears, or *nigigwag,* otters, had seen an Anishinaabeg, and unlike the bineshiinyag and the winged ones, they could not fly above the earth and see what was happening. So, the first thing that Mishi-makwa and Nigig decided to do was to open a line of communication between the beings who live under the earth, the beings who live in the water, and the beings who live on the earth.

"We need to dig a hole so that we can reach the Anishinaabeg on the surface," said Mishi-makwa.

"Oh good," said Nigig, "That is something that we are both really good at. With my claws and your claws, we should be able to dig a tunnel quite nicely."

So they dug, and dug, and dug . . . and dug . . . and dug. When Mishi-makwa would dig too vigorously, rocks would fly through the air and hit Nigig on the head. When Nigig would dig, he found that he was not strong enough to move very much earth, and he only ended up kicking up sand, which got into the eyes of Mishi-makwa.

"This just isn't going to work!" exclaimed Mishi-makwa.

"Oh, you think?!" snapped Nigig, as he rubbed the sore spot on his head where a rock had just hit him. "What we need is a tool. Something that we can push up through the earth to make a hole for us so that we can get up there and find the Anishinaabeg. I am going to go and ask Amik to cut down a nice tree for us to use."

Before Mishi-makwa could voice his opinion, Nigig had slid down his slide and was off to find Amik.

Well, first Amik gave them one of his favorite kinds of wood, the aspen, which the amikwag love to eat and the bark of which is filled with lots of medicine. When Mishi-makwa and Nigig put that piece of aspen in the hole and spun it around, it just broke off right away.

Then they tried all of the different kinds of wood that Amik could find, but none of these woods worked for them. Last of all, they tried the beech, but it was too heavy for Nigig to lift, and even with all his strength, Mishi-makwa could not lift the beech by himself.

So they sat around wondering what to do, when Nigig said, "Wait a second! We're clan animals, and what is the use of being a clan animal if we do not use the powers that Gichi-manidoo gave to us? Why don't we use this power to contact Gichi-manidoo and ask him what we should do?!"

Well, Mishi-makwa thought this over for a while, as he does with all new ideas, and he decided that Nigig was right; they should ask Gichi-manidoo for assistance. After all, Gichi-manidoo created the Anishinaabeg. Mishi-makwa and Nigig put down their asemaa and prayed to Gichi-manidoo. Having never before asked Gichi-manidoo for advice on anything, Mishi-makwa and Nigig were not sure if they would get an answer. They nearly jumped several feet into the air when they heard Gichi-manidoo ask them what they needed. So Nigig explained the situation in great detail, beginning with how Ogiishkimanisii had parted Amik's hair so that he would listen, and Amik being too busy to carry the message, and so on.

Finally, just when Mishi-makwa had almost fallen asleep listening to this long saga that Nigig was telling, Nigig ended by saying, "So, we need a tree that will be strong enough to push through the earth, but light enough to carry, and oily so that it will slip easily through the earth." And he described all of the characteristics of the tree we now call *Nookomis giizhik,* Grandmother cedar.

Because Mishi-makwa and Nigig asked Gichi-manidoo for help in a respectful way and because they wanted help for one of their fellow creations, Gichi-manidoo granted their request. There beside where they stood, a new tree shot up, fully grown. Mishi-makwa and Nigig thanked Gichi-manidoo, picked up the new tree, and began pushing it through the earth.

Meanwhile, above the earth perched *Baambiitaa-binesi,* a sort of woodpecker whom we do not see anymore. Baambiitaa-binesi loved to go up to wood and knock out a message with his beak about anything he saw, sending it to the other winged ones. For instance, if he saw an ant, he would dutifully knock out full descriptions of all of the ant's activities. He sent so many messages about so many minuscule things that the other winged ones tried to ignore him most of the time.

Well, on this day, Baambiitaa-binesi just happened to be in the right place at the right time to see something really interesting happening to the earth, so he flew to the nearest mitig to knock out a message to the bineshiinyag and the other winged ones: "The ground is shaking. The ground is shaking."

The bineshiinyag heard this, and they said to Baambiitaa-binesi, "Well if the ground really is shaking, you'd better get out of there because you know what happens when the ground shakes: KA-BOOM!"

Baambiitaa-binesi knocked out his reply: "No, No, it is different this time. There is no steam. It is not hot at all. I think this time is different. Wait, I'm going in for a closer look."

The other winged ones shook their heads and went about their business, not expecting to hear from their message relayer again any time soon.

Baambiitaa-binesi landed right on the spot where the ground was shaking. He began shaking. As he was bouncing all around, he made as many mental notes as he could about the situation. He felt his feet; they were not hot. He felt his feathers; they had not been singed. He smelled the air. There were no funny odors.

Then he started to feel a little dizzy from all the shaking, so he flew over to the nearest tree to knock out his message to the other bineshiinyag, "The ground is shaking. I am dizzy."

The other bineshiinyag responded, "Yes, we know you are dizzy, Baambiitaa-binesi. It's from all that pounding on trees. It is a wonder you have not pounded your brains out before now."

Ignoring these comments, Baambiitaa-binesi continued, "My feet are not hot. My feathers are not singed. There are no funny smells in the air. I will stay right here and let you know what happens."

"You do that," said the other bineshiinyag. "We are going to fly to safety."

True to his word, Baambiitaa-binesi stayed on his perch. And then the ground gave one last shake, and a giant hole appeared. Just as Baambiitaa-binesi was about to tap this message out to his audience, Nigig jumped out of the hole. So Baambiitaa-binesi began a new message: "There's a hole where the ground was shaking. And now a nigig just jumped out of the hole!"

"Okay, now we know you have hit your head against one too many trees," responded his listeners. "Nigigwag live in the water. They do not dig tunnels and come up through holes in the ground."

But Baambiitaa-binesi was too absorbed in what was happening to pay any attention to them. He kept tapping out new messages, "Something green is coming out. Something green is coming out from where the nigig came out. It's green and soft. I think it's grass. No, it's bigger than grass. It has wood, and branches. It's a tree. It's a new tree."

"A new tree, eh?" responded his listeners, "It's probably purple with yellow polka dots, right?"

Still ignoring them, Baambiitaa-binesi watched the rest of Nookomis giizhik come out of the ground. He saw the green, flat foliage come through, and then the trunk, and the trunk got thicker and thicker as it came further and further out of the ground. And then he saw the roots, more and more roots. Finally, the entire tree, roots and all, was out of the ground.

And then Nigig, who was walking around the tree in circles, when he was supposed to be pulling on the top of the tree while Mishi-makwa pushed from below, went up to the hole, and yelled down to Mishi-makwa, "Just one more push and we should be there." So Mishi-makwa gave one huge push, and the whole tree popped out of the ground. So did Mishi-makwa.

As Baambiitaa-binesi looked at the tree he noticed that the top of the tree was exactly the same shape and in the same proportion as the roots of the tree. As soon as he recovered from the shock of seeing all of this, Baambiitaa-binesi diligently reported to his audience, who by this time were laughing so hard that they nearly fell off of their perches.

Mishi-makwa walked over to Nookomis giizhik, thanked her for her help, picked her up, and planted her back in the ground. Then he began looking for the Anishinaabeg. Not seeing any of them, he called out, "*Aaniin! Boozhoo!* Is anybody there? Can anybody tell us where the Anishinaabeg are?"

Always the eager reporter, Baambiitaa-binesi jumped to attention, flew right over to Mishi-makwa, and told him where to find the Anishinaabeg. Following his instructions, Mishi-makwa and Nigig walked along the top of the dune beside the shore. As they walked, cedar trees sprang up in their footsteps. Soon they found a group of Anishinaabeg on the beach pounding bark.

Upon seeing this, Mishi-makwa said to Nigig, "Now why do you suppose they are just pounding that bark without eating it?"

Nigig replied, "Why would they want to eat bark?"

Seeing a gigantic makwa and a nigig coming toward them, the Anishinaabeg stopped what they were doing and ran in terror, screaming.

"Well," said Mishi-makwa, "What are we supposed to do now? These Anishinaabeg are difficult to find. When we do find them they make awful noises, and they won't stay around long enough for us to talk to them."

Then they heard another sound. This one was even louder than the sound they had heard the Anishinaabeg making as they ran away. Nigig and Mishi-makwa looked at each other, and they both looked at Baambiitaa-binesi, who was just as confused as they were. It was an awful noise, but they all proceeded in the direction from which it came. There, on the other side of one of the sand dunes, was a little *abinoojiiyens,* a baby, who had been left behind in the commotion. Mishi-makwa and Nigig went up to the abinoojiiyens for a closer look, causing the abinoojiiyens to open his mouth really wide and let out one long, bloodcurdling scream.

"Well that's the problem! That's what's wrong with these Anishinaabeg!" exclaimed Mishi-makwa.

"What?!" yelled Nigig, straining to hear over the racket that the abinoojiiyens was making. "What did you say?"

"I know what is wrong with these Anishinaabeg," yelled Mishi-makwa in a great booming voice that overcame the scream of the abinoojiiyens. "See, this Anishinaabe does not have a last berry."

"A last what?"

"A last berry. He does not have a last berry to keep the other berries down. See, I have one," said Mishi-makwa, opening his jaws so that Nigig could see inside to the pink piece of round flesh hanging on the back of his throat.

"Why would anybody want a berry?" said Nigig in disgust. "I think that fish are so much tastier than berries."

"He needs a last berry or he won't have anything to chew on," explained Mishi-makwa.

"Well, if you think this loud little thing needs a last berry," said Nigig, "then give him one and let's get back under the earth. My fur is drying out."

So Mishi-makwa climbed up onto the dune near the water's edge, and slid down, and where he slid a bunch of trailing, wooded vines grew. When he reached the bottom of the cliff, Mishi-makwa stood up, shook off the sand, walked over to the vine, and picked off one red berry. Mishi-makwa took this berry over to the abinoojiiyens, and just as he opened his mouth to scream again, Mishi-makwa dropped in the berry. The abinoojiiyens did not scream. He swallowed. He waited a moment and swallowed again.

Then Mishi-makwa spoke to the abinoojiiyens in a very quiet voice: "There you go. Now you'll be all right. Now you will know what to do."

That is why we see abinoojiiyensag swallowing so much; they are trying to keep their last berry down.

Eventually Mishi-makwa and Nigig were able to find a group of Anishinaabeg and explain to them the wonderful gifts that Nookomis giizhik had to offer them. Then Mishi-makwa and Nigig went back under the earth where all the other beings of Creation were eagerly waiting for them to hear about the Anishinaabeg. Nigig told them how he had saved the Anishinaabeg, with the help of Mishi-makwa, of course.

Mishi-makwa and Nigig saved the Anishinaabeg. They taught the young to swallow their last berry so that they would learn to keep down the rest of their food. They also brought the Anishinaabeg Nookomis giizhik, who opens the line of communication between the worlds, and who has many physical, medicinal, and spiritual virtues that enable the survival and prosperity of the Anishinaabeg.

Mii sa iw. Miigwech. That is it. Thank you.

## NENABOZHO AND THE ANIMIKIIG

One day Nenabozho was seeking fame.[2] He wanted to do something so incredible that the Anishinaabeg would sing songs and tell stories about him. So he went to an anishinaabe village to ask what great task he could accomplish to get the fame he sought. Unfortunately, it was a very busy time when Nenabozho went to the village. And, although they greeted him, none of the adults had any time to stop and visit with him. Nenabozho continued through the village until he came to an area where the children were playing. Upon seeing him, the children dropped their toys and raced over to greet the great Nenabozho.

"Oh children!" exclaimed Nenabozho. "What am I to do?"

The children looked at each other, confused, and then one boy mustered up enough courage to ask, "What's wrong, Nenabozho?"

"Well," said Nenabozho. "I have a problem, and I need your advice. Could you help me?"

2. I have never heard this story labeled as an aadizookaan or a dibaajimowin, but I believe it is an aadizookaan because it is about Nenabozho and contains information of how an important tree came to the Anishinaabeg. This retelling is based on the story that my mother, Mary Geniusz, told me several times as a child. She said that Keewaydinoquay told her this story. Mary Geniusz also writes this story in her manuscript (2005, 58–65). Densmore also provides a written version of this story ([1928] 1974, 381–84).

The children nodded excitedly. The boy exclaimed, "Oh, yes, Nenabozho, we can help you. What do you need?"

"I want to be famous. I want the Anishinaabeg to remember me in songs and stories for years and years to come. What can I do to be honored for generations?" asked Nenabozho.

All the children looked blankly at each other. They shrugged, and then turned expectantly to the boy who had become their spokesman. He thought for a moment, then opened his eyes widely and exclaimed, "You could be a great hunter and bring us lots and lots of meat. Great hunters and the feasts they give are always remembered for years and years." He looked around to the other children for their approval, and they all nodded excitedly.

"That is a marvelous idea!" Nenabozho declared, "I will be a great hunter. I will bring lots of meat to my relatives, the Anishinaabeg. I will make such a great feast that I will be remembered by your grandchildren and your grandchildren's grandchildren." So off he went.

Now there was one problem with this plan. The reason that Nenabozho had never been revered as a great hunter before was that he was not a great hunter. So, when he set off to impress the Anishinaabeg with his hunting skills, he failed miserably. He did not kill even one animal.

"Whew," Nenabozho complained to himself, "hunting is not for me. There has to be an easier way to do this. How am I supposed to be a great hunter if I don't even see an animal?"

And then Nenabozho had a splendid idea. He would ask his father to send him a great wind from the West to blow him a herd of animals. So Nenabozho prayed to his father, and within moments a tremendous wind began to blow from the West. The wind blew and blew, and soon animals began hurtling through the air and landing at Nenabozho's feet. There were beaver, deer, birds, and otters. Nenabozho had to step quickly out of the way as a bear flew by, followed by a buffalo. And then the strange ones started coming. Animals with long noses, stripes, dots, and long necks. The pile of meat grew and grew. Nenabozho smiled with pride; he could just picture groups of Anishinaabeg sitting around winter fires telling stories about this magnificent day when he, Nenabozho, had brought them all of this meat. When the pile was so tall that Nenabozho could not see the top of it, the wind stopped. Nenabozho brushed off the dirt, which had blown toward him along with the animals, and set off to invite the Anishinaabeg to a feast.

When the Anishinaabeg saw a gigantic pile of meat just outside of their village, they were not happy at all. Instead of praising his hunting skills, they complained that they were going to have to move the village when all of that meat began to rot. Instead of marveling at the strange animals that he had killed, the Anishinaabeg complained that they could not even use these strange striped and dotted hides.

They left to pack up the village, and Nenabozho sat down beside his kill and shook his head. Wasn't there anything he could do to become famous? Just then the boy who had suggested the hunting idea in the first place came over to Nenabozho and said, "Maybe it isn't just the amount of animals a hunter brings to the village; maybe it's also the stories that the hunter tells about how he hunted those animals."

"Stories about hunting, *na?*" said Nenabozho. He thought for a moment and then jumped up. "I know; I could kill the most famous animal and bring it back to the village and tell the Anishinaabeg all about my hunting adventure. Then they will tell my story for years to come. Who is the most famous animal?"

"That's easy," said the boy. "*Name,* the Great Sturgeon. Anyone who could kill Name would be famous indeed."

Nenabozho thanked the young man, and then raced off to find his grandmother. She would surely know how to defeat the great Name.

Nenabozho's grandmother was not eager to help him in his new pursuit. She thought it was too dangerous, but after a great deal of persuasion, she finally said to her grandson, "If you want to kill Name, you have to use an arrow fletched with the feathers of an *animikii,* thunderbird."

So Nenabozho set off to kill an animikii. He walked to the west until he came to the base of the great mountains where the animikiig make their nests. He saw one of these nests very high atop one of the mountain peaks. Now all he had to do was think of a way to get into that nest. He watched as two animikiig left the nest and then returned, carrying small animals in their talons. "I know," he exclaimed quietly to himself. "I will turn myself into a rabbit, and they will think I am a good meal." So he turned himself into a rabbit, and before long he was flying through the air in the talons of the father animikii, who deposited him in a nest filled with baby animikiig. The mother soon came back to the nest, and said to her husband, "What are you doing? Why would you give our babies that rabbit? Don't you know that Nenabozho is in the area? We must be very careful."

"Don't worry," said the father, "it's only a little rabbit. Our children are safe." And off they flew to find their own dinner.

As soon as they were gone, Nenabozho changed back into himself and killed the baby animikiig. Then he quickly scooped up a handful of their feathers and began to run.

Although she was far from the nest, the mother animikii suddenly knew that her babies were gone. The two animikiig knew Nenabozho had tricked them, and they set off to find him, which was not difficult because he was proceeding rather slowly down that steep mountain.

As soon as they saw Nenabozho, the animikiig began throwing bolts of lightening at him. Nenabozho jumped to the side to avoid one bolt, just to have another graze his foot. He tried hiding behind an *ishkaatig,* a maple, only to have the animikiig split the tree in half. Nenabozho kept running. Bolts of lightening broke through the darkness as night fell. He ran down the mountain, leaving a path of smoldering trees behind him.

Just as he was getting very tired and sure that he could not jump out of the way of a lightening bolt one more time, Nenabozho came upon a grove of trees with white bark. A hollow log from one of these trees lay on the ground directly in his path, and he quickly dove into it.

Seeing their target attempting to hide from them yet again, the animikiig flew directly toward the hollow log. And then they stopped. Inside the log, Nenabozho lay huddled in a ball, not making a sound for fear the animikiig might hear him. He just lay there, afraid that even his breathing would be too loud. Then he heard the animikiig.

"You have won, Nenabozho. You have found the one place to hide where we will not strike, inside the bark of a King-child. This is our own child, and we will not strike this tree."

Not sure if he was being tricked, Nenabozho lay inside the hollow log until he was certain that the animikiig had left the area, then he slowly crept out of the hollow white log. "*Howa!*" he said to himself. "This tree that sheltered me is the child of the animikiig. This tree will be very useful to my nephews the Anishinaabeg because they can stand under it in a storm and not be struck by the animikiig. If they use the bark of this tree to make their lodges, they will also be safe in any storm. Since this bark lasts for such a long time, even longer than the tree inside of it, the Anishinaabeg can use it to protect their things."

So Nenabozho marked the bark of this tree, *Nimishoomis wiigwaas,* Grandfather birch, with little black lines in the shape of the baby animikiig. These markings would show the Anishinaabeg where to find this tree, and remind the animikiig that this was their child.

And then off went Nenabozho, in search of Name. Mii sa iw. Miigwech.

These aadizookaanan and dibaajimowinan carry information about inaadiziwin as well as pieces of gikendaasowin about giizhikaatig and wiigwaasi-mitig. I have used these aadizookaanan and dibaajimowinan as a guide for conducting new research with elders and decolonizing existing research on gikendaasowin about these trees. These aadizookaanan and dibaajimowinan, especially "The Cedar Song," provide the framework through which I present this research in the rest of this chapter.

Also included in this chapter are references to many items that the Anishinaabeg make from giizhikaatig and wiigwaasi-mitig. When I first explained my project to Dora Dorothy Whipple, she said that this text would be an opportunity to teach people how the Anishinaabeg make things rather than just telling them that the Anishinaabeg do make things. She criticized other texts for only listing items that could be made rather than giving specific instructions (Whipple, pers. comm.). This chapter is meant to be a sample of decolonized gikendaasowin, and hopefully a resource for cultural revitalization programs, so I thought it was important to include instructions on how to make the items mentioned in this chapter. These instructions can be found in the appendix.

## GIWAABMAA GIIZHIK: BEHOLD THE CEDAR

The first Ojibwe line of "The Cedar Song" is "giwaabamaa giizhik"; in the English verse, this line is "behold the cedar," which is very close to the literal translation of the Ojibwe line: "You see cedar." When learning this song, one learns one of the Ojibwe names for this tree: giizhik. This happens to be one tree that many Anishinaabeg can name in Ojibwe, but if it were not, this would be a good time to check the literature on gikendaasowin and to ask Ojibwe speakers to see if this or other names for this tree are found anywhere else. When doing this search, we find that this name is found other places. Huron Smith, for example, writes the Ojibwe name for this tree "gi'jîg" (1932, 421). Meeker, Elias, and Heim have "giizhik" and "gizhikens," a diminutive of giizhik (1993a, 387). Ojibwe speakers use this name, along with some variations, including "giizhikaatig."

We know from the dibaajimowin "Traditional Anishinaabe Advice to Youth," that giizhikaatig is associated with the birch, and therefore it makes sense to teach about these trees together. We have ways to refer to the cedar in Ojibwe, but what about the birch? By going through the colonized literature and checking with Ojibwe speakers, we find that the Ojibwe names for birch are "wiigwaas," "wiigwaasi-mitig," and "wiigwaasaatig."

## MITIG AZHITWAA: THE HOLY TREE

The next Ojibwe line of "The Cedar Song" is "mitig azhitwaa," and the corresponding English verse is "the holy tree," which appears to be an accurate translation of this line. "Mitig" means "tree," and "azhitwaa" probably comes from the verb *izhitwaa,* meaning "to be holy." By asking elders and looking at written sources we can explain the importance of both giizhikaatig and wiigwaasi-mitig to inaadiziwin and izhitwaawin. As Whipple says, "Anooj igo inaabadizi aw giizhikaandag" (cedar has many uses) (pers. comm.).[3] Keewaydinoquay makes a similar statement about giizhikaatig: "So many good things this tree does for us, physical healing, spiritual healing, physical protection, spiritual protection. They are so many that you wouldn't believe it if you didn't see them" (Keewaydinoquay 1986).

Densmore says that these giizhikaatig and wiigwaasi-mitig are considered "sacred" to the Anishinaabeg because of their "great usefulness" and because

3. Whipple always translates "giizhikaandag" as "cedar," but it could also refer specifically to cedar boughs.

they are both linked to Nenabozho ([1928] 1974, 381). In her text, Densmore presents two aadizookaanan, told to her by Mary Razer, about Nenabozho's association with both of these trees. One of these aadizookaanan, which I call "Nenabozho and the Animikiig," was retold earlier in this chapter. The other aadizookaan she presents is about six Anishinaabe men who go in search of Nenabozho to ask him how they can restore to life one of the men's daughters. When they find Nenabozho, he has a giizhikaatig growing out of his head. They follow his instructions, the final one of which is to lay the spirit of the girl, tied into a bag, on a bed of giizhikaandag (cedar boughs) in a sweatlodge, and the girl comes back to life ([1928] 1974, 384–86). Huron Smith also describes the importance of giizhikaatig: "The Ojibwe worships the Arbor Vitae or White Cedar and the Paper or Canoe Birch, as the two most useful trees in the forest. The pungent fragrance of the leaves and wood of the Arbor Vitae are always an acceptable incense to Winabojo" (1932, 421–22).

Both of these trees have important spiritual properties. Keewaydinoquay teaches that giizhikaatig is the most sacred tree to all of the Great Lakes peoples (1988). As the aadizookaan "The Creation of Nookomis Giizhik" explains, giizhikaatig connects the worlds and opens our communication with the manidoog. According to the dibaajimowin of Keewaydinoquay, the growth pattern of giizhikaatig shows this connection because this tree exists in the same form both above and below the ground. For every branch that we see above the ground, there is a root below the ground, and for every piece of foliage that we see above the ground, there is a rootlet below the ground. Birds nest in the branches of giizhikaatig and small animals, such as mice and rabbits, nest in the roots of this tree. This structure of giizhikaatig symbolizes a dibaajimowin integral to inaadiziwin: the importance of living in balance (Keewaydinoquay 1986; M. Geniusz, pers. comm.).

Keewaydinoquay also says that giizhikaatig is one of the two most important ingredients in asemaa, or kinnikinnick, the other being makwa-miskomin (bearberry, *Arctostaphylos uva-ursi* [L.] Spreng.). We know from the aadizookaan "The Creation of Nookomis Giizhik" that these plants were created to help the Anishinaabeg, and that Mishi-makwa, the Great Bear, helped to bring both of these plants to the Anishinaabeg. Although the dibaajimowin about what herbs should be included is different in different anishinaabe communities, Keewaydinoquay says that giizhikaatig and makwa-miskomin are always present in kinnikinnick. She adds that a kinnikinnick that will be smoked in a pipe should contain giizhikaatig, but only a small amount because giizhikaatig, like all of the conifers, has resins in it that give a harsh taste to the smoke (1986).

Wiigwaasi-mitig also has spiritual properties, and this tree was given to us by the manidoog, as explained in the aadizookaan "Nenabozho and the Animikiig." Wiigwaas, birch bark, has an important part in ceremonies. As mentioned in chapter 2, Midewaajimowin are often recorded on scrolls of wiigwaas, as are ceremonial instructions, healing songs, and medicinal recipes. According to Angeline Williams, other spiritual customs associated with wiigwaasi-mitig include putting the dried root of a "wild pea" inside the "thin inner bark of a piece of wiigwaas,"[4] and then tying it with basswood and wrapping it with a piece of buckskin so that it will not be lost. The person who carries this will have good luck with anything for which she or he asks (EWV, notebook 19b). Williams also describes bitten wiigwaas patterns being used "like fortune telling," for activities such as hunting.[5] The different symbols, such as bows and animals, foretell good luck, but there are also symbols that foretell bad luck. If the prophecy is bad, the person whose fortune is being told generally throws the patterns of wiigwaas away. If, on the other hand, the prophecy is good, that person keeps the patterns and uses them "as wrapping," presumably referring to the wrapping around the dried root of the wild pea (EWV, notebook 19b). From wiigwaasi-mitig we also get dishes, in which certain items are placed and burned or laid out in the woods during a ceremony. This is a good alternative to using Styrofoam or plastic dishes, which are dangerous to burn and harmful to the environment if left in the woods.

### ONESENODAWAAN BIMAADIZIWIN: COME LET HER BREATHE YOU NEW LIFE WITHIN

"The Cedar Song" continues with the Ojibwe line "onesenodaawaan bimaadiziwin"; in the English portion this line is "come let her breathe you

4. Williams adds that sometimes this inner bark is "bitten" so that it looks like the head of a person (EWV, notebook 19b).

5. Williams adds that "those with good teeth did this" (EWV, notebook 19b).

new life within." The literal translation of the Ojibwe line is "she breathes life into them [the Anishinaabeg]." When speaking to elders and looking at the literature, these lines become clearer, and they can be applied to giizhikaatig and wiigwaasi-mitig because both of these trees bring spiritual and physical life to the Anishinaabeg. For example, in the story presented by Densmore and summarized in the previous section, we see an example of giizhikaatig's bringing a girl back to life ([1928] 1974, 384–86).

From giizhikaatig comes an oil that the Anishinaabeg use in many ceremonies and important times of life. In the *madoodiswan* (sweatlodge), Keewaydinoquay says, a person should cover his or her entire body, from the top of the head to the bottom of the feet, with cedar oil. She explains that practice gives a person "wonderful vibrations and many blessings. And it also makes it so no matter how hot the steam is, nobody's ever burned" (1986). Two of Keewaydinoquay's oshkaabewisag conducted my marriage ceremony. During this ceremony, my husband and I were covered in cedar oil, and giizhikaandag (cedar boughs) hung over the place where we were married. Cedar oil is also used whenever someone has a hard time in his or her life, or needs, as Keewaydinoquay describes it, "a little extra boost." A person should use cedar oil on a daily basis by putting a drop of cedar oil on his or her finger and touching the center of his or her forehead and "life spot" with that oil. The life spot is the fleshy part on the base of the neck. While doing this the person says, "I give thanks for the gift of life, and I unite myself with the ongoing of the people" (Keewaydinoquay 1989a). Keewaydinoquay describes the "many miracles" she has seen result from cedar oil. Some individuals to whom she had given the dibaajimowin about how to anoint oneself with cedar oil tried it, and it prevented them from committing suicide because they realized, as they were doing this, that they were saying thank you for life at the same time that they were thinking of ending their lives. Keewaydinoquay says that she has also seen cedar oil bring a child back to life that was not breathing (1989a).

In izhitwaawin, giizhikaatig and wiigwaasi-mitig are both present at the start of a new life. Keewaydinoquay says that when a child is born one of those attending the birth covers his or her hands in cedar oil and cups those hands over the newborn's nose and mouth so that the baby will smell the cedar oil and be encouraged to breathe. She adds that this is a much more pleasant way to come into the world than with a smack to encourage the baby to breathe, as is done in so many hospitals today (1986). Whipple remembers, when she was about five years old, being present when her aunt was having a baby and seeing the floor covered with ***mashkosiwan*** (hay or grass), and seeing giizhikaandag there as well (pers. comm.). Hilger describes Anishinaabe babies having their first bath in wiigwaas: "A baby's bathtub was a nonleakable birchbark receptacle . . . Its length was the distance from finger tips to elbow; its width two stretches of one hand" ([1951] 1992, 18).[6]

In izhitwaawin, giizhikaatig and wiigwaasi-mitig continue to surround a child after birth. When a child is named, every opening of his or her body is covered with cedar oil so that nothing bad should ever enter that child (Keewaydinoquay 1986). Anishinaabeg make the ***dikinaagan*** (cradleboard) out of giizhikaatig (Keewaydinoquay 1986; M. Geniusz, pers. comm.). Geyshick adds that the dikinaagan can also be made out of birch (1989, 22). Babies are strapped into a ***waapijipizon*** (moss bag) before being tied into the dikinaagan. There is a tray made of wiigwaas, called an ***apabizon***,[7] inside of this moss bag that holds both the baby and the ***aasaakamig*** (the moss used to diaper the baby) (Rose, pers. comm.; M. Geniusz, pers. comm.). Through their infancy, Anishinaabe children are held in this way by giizhikaatig and wiigwaasi-mitig.

Giizhikaatig contains many life-sustaining medicines, including medicinal properties, which help the body to heal and maintain good health. Keewaydinoquay teaches her students that giizhikaatig has a high content of vitamin C, which is very important to overall health (Keewaydinoquay 1988; M. Geniusz 2005, 47).[8] Giizhikaatig is one ingredient in a cough syrup

6. Hilger also describes the bathwater used for this first bath: "Certain plants, among them catnip, spruce boughs, and twigs of gībāīmīnā'mīgōk, a plant that grows in swamps, were boiled in water and the baby bathed in the decoction immediately after birth or at least on the day of birth. Some used 'an herb that can be found where maples grow'. . . . The bath was thought to give the child a strong constitution and it might be repeated by the mother any time she desired" (Hilger [1951] 1992, 18).

7. See the appendix for more information on the apabizon, the birch bark tray inside of the waapijipizon.

8. Keewaydinoquay teaches that all of the conifers have a high vitamin C content, and any of them can be used as a source of vitamin C, which is often administered as a tea

recipe (Densmore [1928] 1974, 340–31).[9] The Anishinaabeg use giizhikaatig to treat colds by making a tea from the foliage of this tree (M. Geniusz, pers. comm.).[10] This tea also settles stomach gas (M. Geniusz 2005, 47), and it is good for taking the "fishy" taste out of one's mouth after a fish dinner (Keewaydinoquay 1986).

The Anishinaabeg also use giizhikaatig to sustain life by using this tree to disinfect an area where there has been sickness. Keewaydinoquay says, "A kinnikinnick that's made very strong with . . . grandmother cedar can be used as a fumigant in a sick room." (1986). Gilmore writes that twigs of giizhikaatig are burned to fumigate a home in which someone was sick with a contagious disease. He continues, as mentioned in chapter 1, "Many years ago when smallpox first came to the Chippewas they moved into arbor vitae swamps and camped there during the period of the plague" (1933, 123). Mary Geniusz elaborates further on how moving into a swamp filled with giizhikaatig would have helped these Anishinaabeg during such an epidemic. She says that these Anishinaabeg were trying to get as strong a medicine as possible to fight the epidemic. By living among giizhikaatig, they were able to have constant access to both her physical and spiritual properties. They would be walking on the fallen foliage, causing it to release vapors, which are, as Gilmore points out, a disinfectant. It is possible they were burning the foliage to release more curing properties into the air. They could have eaten the foliage straight as well, which would have given them internal doses of medicine. Geniusz says that living among giizhikaatig would also have provided these Anishinaabeg with spiritual help as well "because it is through cedar that you call for help. . . . You can just put your hand on the tree itself and say 'Grandmother help me' and it

---

(Keewaydinoquay 1988; M. Geniusz 2005, 47). A very strong cedar tea can cause an abortion; therefore, pregnant women should not drink it. One should be aware, also, that certain conifers have been introduced as ornamental shrubs. Some of these are poisonous and should not be used as a source of vitamin C. Make sure to properly identify the conifer as nonpoisonous before drinking a tea made from it.

9. Densmore does not give the rest of this cough syrup recipe (Densmore [1928] 1974, 340–41).

10. See the appendix for instructions on making cedar tea.

will go straight into the sprit world because that is what cedar does" (M. Geniusz, pers. comm.).

Giizhikaatig and wiigwaasi-mitig can sustain life, as described in Keewaydinoquay's dibaajimowin "Traditional Anishinaabe Advice to Children," by providing a person with everything he or she will need to survive: water, food, fire, shelter, transportation, and medicine. What follows is a brief description of how these trees can do this.

Keewaydinoquay told Mary Geniusz that giizhikaatig likes to grow where "her feet can be in water," so where one finds this mitig, one will also find a water source (M. Geniusz, pers. comm.). Giizhikaatig and wiigwaasi-mitig can provide us with food because the inner layer of the inner bark, or cambium,[11] of both giizhikaatig and wiigwaasi-mitig provide an emergency food. On giizhikaatig, this is the same light-colored inner bark that is used to make cordage and mats (M. Geniusz, pers. comm.; Keewaydinoquay 1988). The inner bark of wiigwaasi-mitig is harder to describe because the bark from this tree peels off in so many layers. This edible part is, however, the part left on the tree after taking a piece of wiigwaas off the tree to make an item such as a *makak* (birch bark basket) or a *wiigwaasi-jiimaan* (birch bark canoe). Mary Geniusz says that this inner cambium on the wiigwaasi-mitig is quite distinguishable from the wood of the tree and that when it is not the right time of year to collect wiigwaas, as will be discussed later, this inner cambium sticks to the birch bark as it is peeled and has to be separated afterward (pers. comm.). The inner cambium of both of these trees is a very starchy food, which can be eaten straight or cooked like spaghetti. Of the two trees, wiigwaasi-mitig has a sweeter taste; some native people, including some Anishinaabeg, make syrup out of the sap of this tree. (M. Geniusz, pers.

11. The inner bark and the cambium are the same thing. James Underwood Crockett describes the cambium as "Only one cell thick" and says it is under the phloem, which is under the outer bark, the part of the tree that we can see without peeling away any bark (1972, 20). Lee Peterson describes "the inner bark (cambium)" of trees being used as flour (Peterson 1977, 295). Keewaydinoquay uses the word "inner cambium" when describing this emergency food source (1988). I have made and eaten bread made out of the light-colored inner bark of balsam fir, which like cedar is a conifer and is also said to have edible inner bark, and this lighter inner layer of the inner bark is very pliable, chewable, and grindable, whereas the darker outer layer of the inner bark acts more woody, and is not as easy to chew or grind.

comm.; Keewaydinoquay 1988). This starchy cambium can sustain a person until he or she has the energy to make other tools to get different kinds of food, and these trees can help in the making of these tools as well.

The inner bark of giizhikaatig also makes strong cordage, which can be used for a number of different things, including making tools used to get food, such as fishing lines, snares, or bowstrings. Mary Geniusz writes that a fishing line made out of cedar rope might not hold a fish as big as a pike, but it would hold a smaller fish. A bowstring made of cedar would only last for a few pulls, but if one knows how to hunt with a bow, this would be enough to kill a deer (M. Geniusz 2005, 47). Although there are other materials that will make a stronger cordage, the process of making cedar bark cordage is not as hard on the hands as the process of making cordage out of other materials, such as the nettle (*Urtica dioica* spp. *gracilis*) (Keewaydinoquay 1986; M. Geniusz 2005, 47).[12] Keewaydinoquay describes the basic process of making cordage from the inner cambium of giizhikaatig, saying, "These long fibers are rolled between the fingers and spun against the leg, and it makes a wonderful cordage that can be used for all kinds of things" (1986).[13]

From wiigwaasi-mitig we can make dishes for gathering, preparing, and cooking food. George McGeshick, Sr., says that, as a child, he used to eat *ziinzibaakwad* (maple sugar) out of a cone of rolled wiigwaas (pers. comm.). Kathryn Osogwin explains that a *biskitenaagan* (a folded container of wiigwaas usually used to collect sap) can be used to boil liquids in an emergency situation. One puts a handle on the sap collector before using it this way

12. There are several species and subspecies within the genus of *Urtica* L., and, according to the U.S. Department of Agriculture, some of these are considered "native" and some are considered "introduced" (USDA, NRCS 2006). Daniel E. Moerman lists this particular subspecies of nettles, *Urtica dioica* spp. *gracilis,* as being used by the Anishinaabeg and tribes near them for cordage (1998, 580). It is of course possible that the Anishinaabeg use other subspecies of nettles as well. Mary Geniusz describes how to make cord from nettles. First one hangs the nettles until they are dry, because once the plant is dry, the "stinging hairs" on the plant will no longer sting. Then one breaks down the stalks. The outer, woody layer breaks away when it is dry, leaving one with just the fiber. This fiber should be soaked before spinning it into a cord. Geniusz recommends spinning them against one's legs to make a cord (2005, 47).

13. See the appendix for more information on gathering the inner bark of giizhikaatig and using it to make rope.

(EWV, notebook 20; Nichols and Nyholm 1995, 240). Osogwin also says that water can be boiled in a pail made of wiigwaas, and that to do this the pail is put directly over the fire (EWV, notebook 20). Florina Denomie, of the Bad River Reservation, says that the Anishinaabeg put buckets made of wiigwaas on sticks and lay them over a smoldering fire. She writes that the liquid inside the buckets prevented them from burning. Denomie remembers her grandmother heating water, making tea, and cooking food in these buckets at the sugar bush (WPA 1936–40, envelope 8, 42). Osogwin describes using wiigwaas to cook fish by wrapping an unscaled whitefish in a piece of wiigwaas and laying the whole bundle in hot ashes. When the wiigwaas turns brown, the fish is cooked, and one removes the fish, leaving the skin on the wiigwaas (EWV, notebook 20).

Storage containers for food and other items are also made from giizhikaatig and wiigwaasi-mitig. Keewaydinoquay says that containers made of the bark of giizhikaatig will help prevent the items in those containers from molding, and that for this reason the Anishinaabeg have used such containers to store food and items made out of leather. One can also pack important items away in cedar and sweet fern foliage, and the combination of those two plants would prevent molds from growing and make the stored items smell nice (Keewaydinoquay 1986; 1989a). Osogwin describes storing "dried pounded meat" in a covered wiigwaas container, sealed with pine pitch.[14] She says that food is also stored in storage pits that are covered with a layer of wiigwaas, then a layer of "cedar bark," and then earth. She adds that jam, dried meat, dried fish, whole squash, maple sugar, maple syrup, and cakes made of maple sugar can all be stored in the ground this way, but corn and "other things that couldn't stand moisture or mold," such as dried squash or "rice," are stored in an *ataasoowigamig* (a peaked storage house), the roof and sides of which are covered with bark, used by everyone in the village (EMV, notebook 20). Bags woven from the bark of giizhikaatig are also used to store food. Osogwin says that cedar bags are lined with thin sheets of wiigwaas "like waxed paper," and then food is put between the sheets of wiigwaas "so cedar odor wouldn't taste in the food" (EWV, notebook 20).

14. See the appendix for more information on storing items in a container made of wiigwaas.

Giizhikaatig and wiigwaasi-mitig also provide us with fire, a necessary source of warmth and a tool on which we can cook. A bow drill is often made out of giizhikaatig, and this is one method used to start fires. When I was still in elementary school, I went to a ceremony at a woman's house in Milwaukee. Keewaydinoquay was running the ceremony, and because we were indoors, she chose to use candles instead of fires for the ceremony. When it came time to light the candles, someone asked if anyone had a light. A man announced that he did, and he pulled out a bow drill and a piece of wood with a hole in it. He placed this piece of wood on the woman's living room shag rug carpet. He inserted a stick into the hole in the wood, and then wrapped the bow around the stick. He then began to move the bow back and forth very quickly, causing the stick to move inside the hole in a rapid circular motion. Before long there was smoke coming from the hole in the log. The woman whose house it was began to pale, but my mother assured her that the man was an expert and that both her carpet and dwelling were safe. My mother later admitted that she really did not know the man who was starting this fire and that she too was concerned that the house was about to go up in flames. When a coal was formed in the base of the drill, the man used it to light the candles for the ceremony. The house did not burn down, and I learned that, contrary to my schoolteachers' Indian curriculum, making fire was much more difficult than simply rubbing two sticks together. When I was older, my parents took a workshop where the teacher gave all the students cedar logs and then proceeded to teach the class how to turn their cedar logs into bow drills. My father, Robert Geniusz, has successfully started several fires this way, but my mother insists he start all of them outside of the house and not on her carpet. Mary Geniusz describes how to make fire out of giizhikaatig, saying, "The best fire drills are made of cedar. One makes a hole in a small V-shape in a flat piece of cedar. Then one spins an upright piece of cedar in the hole by means of a bow" (2005, 47–48). Radin also mentions the Anishinaabeg using giizhikaatig to make fire, but he says that they use the stick from the "beech tree," and twirl this stick very fast inside a hole, made inside a piece of cedar (1928, 662). Once one has made a spark with the bow drill, one uses a piece of cedar rope to catch the spark and light the wiigwaas, which will then light the fire. Anyone who has tried to start a log by just burning paper knows that other fire starters generally need a lot of kindling to start a log on fire. Wiigwaas does not. Wiigwaasi-mitig helps us with fire because, as the aadizookaan "Nenabozho and the Animikiig" tells us, this tree was given to us by the animikiig (M. Geniusz, pers. comm.).

Once the fire is started, a small piece of smoldering cedar rope or a piece of cedar bark can be used as an *ishkodekaan* (something that can easily transport fire to another location) so that the labor of using the bow drill and log will not have to be repeated. Before the introduction of modern conveniences like matches and lighters, cedar bark was used to transport fire from one place to another, especially when moving to a new camp. One way this was done was by placing a smoldering cedar rope inside a shelf fungus or between two shelf fungi (M. Geniusz, pers. comm.; Keewaydinoquay 1998, 3).[15] Marie Livingston, from the Bad River Reservation in Wisconsin, also describes transporting fire: "They used dry cedar bark which was frictioned into a fleecy cotton substance in which to carry the fire." Livingston says that the Anishinaabeg tie this "fleecy cotton substance" into "bundles about three inches in diameter and twelve or fifteen inches in length" and tie the whole bundle with *wiigob* (basswood fiber). The hot coals are placed inside this bundle. When they need to make a fire, the Anishinaabeg untie this bundle, put the hot coals down, and pile kindling on top of them, and, as Livingston says, "in a short time a vigorous fire would be produced" (WPA 1936–40, envelope 8, 39).

15. Keewaydinoquay identifies the fungi fire carriers as "any of the group *Xylaria*," and says that they are commonly called "Dead men's digits" (Peschel [1978] 1998, 5, 65). Keewaydinoquay quotes her grandfather describing the process of using this fungi to transport hot coals: "There is really nothing to it. Just set three coals on a dry shelf fungus until they have burned their way into the soft punk a little. Turn another fungus upside down over the top of it and you can carry fire safely through all kinds of travel and wet weather" (as cited in 1978, 3). In this section Keewaydinoquay identifies the fungi as "*JibiEpushKwaegun* (polyporus shelf fungi)," and in the glossary she identifies this Ojibwe name as "Any of the group *Xylaria*" (1978, 3, 65). However, *Xylaria* are not shelf fungi, and Mary Geniusz says that they are not the same plants that Keewaydinoquay showed her when describing the fire carriers. Geniusz says that Keewaydinoquay identified the fire carrier as *Fomes fomentarius,* a hoof-shaped fungus that grows on birch trees. Keewaydinoquay told Geniusz that the Anishinaabeg used this fungus to transport smoldering pieces of cedar rope (M. Geniusz, pers. comm.). When lecturing, Keewaydinoquay mentions carrying fire on a piece of cedar rope, and her slide collection does contain a picture of a smoldering cedar rope (1989a).

Fire made from giizhikaatig is also used to find a lost person. Angeline Williams says that those in her community knew that if they were lost, they were to climb a tree and look around the area. Those looking for that person would start a big fire, and the person who was lost would see this fire from the tree they had climbed and know where the rest of the group was. First the Anishinaabeg would make a big fire, and then they would put lots of cedar on it so that it became a really smoky fire, which would be easier for the lost person to see (EWV, notebook 1941, 19a).

Giizhikaatig and wiigwaasi-mitig provide the Anishinaabeg with torches for hunting and fishing in the dark.[16] Joe Wilson, of the Bad River Reservation, told Peter Marksman, who worked on the WPA Indian Research Project, that the Anishinaabeg make torches out of "cedar bark, birch bark, hazel brush, or basswood timber" (WPA 1936–40, envelope 8, 37).[17] Describing two kinds of torches, George McGeshick, Sr., says the *waaswaagan* is a handheld torch and the *zaka'aagan* is a "headlight." Among other things, the waaswaagan is used when spear fishing in the dark, and the *zaka'aagan* is worn on the head when hunting deer in the dark (McGeshick, pers. comm.).[18]

Before modern housing, many Anishinaabeg lived in a *wiigiwaam* or *waaginogaan*, both of which are often made of giizhikaatig and wiigwaasi-mitig.[19] The wiigiwaam is a peaked or cone-shaped lodge,[20] which looks similar to what one calls in English a "teepee" (Rose, pers. comm.). The

16. For photographs of torches, see Densmore [1929] 1979, plate 56.

17. See the appendix for more information on making torches.

18. McGeshick says that two of the reservations in Wisconsin are named for these torches. Waaswaaganing is the Ojibwe name for Lac du Flambeau, and Zaka'aaganing is the Ojibwe name for Mole Lake, the reservation at which McGeshick is enrolled (McGeshick, pers. comm.).

19. Of course, not every wiigiwaam or waaginogaan is made from or covered with these two mitigoog. McGeshick says that, as with anything made from natural fibers, the Anishinaabeg use the materials that are available to make the waaginigaan (McGeshick, pers. comm.). Osogwin seems to agree, as she told Wheeler-Voegelin that the top of the waaginogaan is usually covered with the bark of giizhikaatig, but if that is not available then elm or basswood bark is used (EMV, notebook 20).

20. It should be mentioned that wiigiwaam is also a general term for any of these houses.

waaginogaan or *waaginigaan* is a domed lodge that one often refers to in English as a "wigwam" (McGeshick, pers. comm.; Nichols and Nyholm 1995, 116). Both of these lodges often have coverings made from either or both wiigwaasi-mitig and giizhikaatig. Osogwin explains that the wiigiwaam usually has wiigwaas on the top and the bark of giizhikaatig on the bottom, but if one is traveling and only intending to use the wiigiwaam for one or two nights, only the wiigwaas coverings are used (EWV, notebook 20). The mats found on the lower portion of the waaginogaan and on the inside of this lodge are often made of the bark of giizhikaatig. Geniusz says that, although mats of either cedar or cattails are often used on the lower portion of the waaginogaan,[21] cedar mats are a lot stronger and last longer than their cattail counterparts (pers. comm.). Keewaydinoquay says that cedar mats are often used on the floor of the waaginogaan because they have a natural deodorant in them, they are very tough, they are easily cleaned when shaken, and they retard mold (1986). Densmore writes that in northern Minnesota, where rushes are not as available, the Anishinaabeg make cedar mats for the floors of their waaginogaan ([1929] 1979, 156). Wiigwaasi-mitig is also used to cover the waaginigaan. *Wiigwaasibak,* or *wiigwaasabakway,* are large sheet-like coverings made of wiigwaas, which can be taken off one lodge, rolled, and transported to a new location. Wiigwaasibak often cover both the wiigiwaam and the waaginogaan (Rose, pers. comm.).[22]

It should be mentioned that the wiigiwaam and the waaginogaan are not just shelters; they are an important part of inaadiziwin. Rose says that where she lives in Canada the Anishinaabeg have built an "Elders' Cabin," in which elders talk to young people. This cabin has no floor, just earth. Rose says that the idea for the Elders' Cabin was inspired by the wiigiwaam because when the Anishinaabeg lived only in such lodges, they always touched *Gimaamaanaan* (Our Mother, the Earth), but now the Anishinaabeg are always up high, not touching the ground. The elders working with the Elders' Cabin think that this disconnection from the Earth may be the cause of some of the problems that the Anishinaabeg are facing today (Rose, pers. comm.).

21. See the appendix for directions on making mats out of giizhikaatig.

22. See the appendix for more information on making a wiigwaasibak (a covering for a lodge).

Giizhikaatig and wiigwaasi-mitig sustain the Anishinaabeg by providing them with transportation in the form of the wiigwaasi-jiimaan (the birch bark canoe). The ribs of the wiigwaasi-jiimaan are all made out of giizhikaatig (cedar wood) and the shell of the canoe is made out of wiigwaas (birch bark). Huron Smith says that the Anishinaabeg prefer to use cedar to make "light, strong straight-grained canoe frames and ribs" (1932, 421–22). Keewaydinoquay says that the Anishinaabeg prefer to use giizhikaatig to make wiigwaasi-jiimaanan because this wood does not decay in water; it is tough, lightweight, and can bend without cracking (1986).

George McGeshick, Sr., has made several wiigwaasi-jiimaanan, including one that is now at the Smithsonian Institution. McGeshick says that the materials to make a wiigwaasi-jiimaan are becoming harder and harder to find around his home on the eastern end of the Michigan-Wisconsin border. In the summer of 2004, my father and I worked with McGeshick and several members of the Chicaugon Chippewa to build a wiigwaasi-jiimaan. Unfortunately, we were not able to complete our project, as we had trouble finding usable giizhikaatig and wiigwaas, a problem that McGeshick says he has encountered for many years now. While we were looking for our materials, McGeshick was able to teach us important details about what materials will and will not work to make a wiigwaasi-jiimaan. The wood of giizhikaatig is used to make the entire frame of the wiigwaasi-jiimaan. This wood must come from a straight tree that is tall enough to have a long, smooth lower portion free of branches. Lower branches, as we discovered, cause too many knots in the wood and make it impossible to split into straight pieces. If the tree is not straight enough, it will not split into straight sheets. Because the giizhikaatig needs to be spilt into very thin pieces, having wood that will split straight is essential (McGeshick, pers. comm.). Unfortunately, the woods in northern Wisconsin no longer have large enough quantities of giizhikaatig and wiigwaasi-mitig to find the right size and shape of each of these trees to make canoes.

### GAA-NOOJIMOWAAD ANISHINAABEG: WE CALL HER SAVING TREE, SHE SAVES THE PEOPLE

The next Ojibwe line of "The Cedar Song" tells us "gaa-noojimowaad anishinaabeg," corresponding to the English line "we call her saving tree, she saves the people." The literal translation of the Ojibwe line refers to the Anishinaabeg, saying that they are "the ones who are healed or cured." As the aadizookaan "The Creation of Nookomis Giizhik" tells us, Mishi-makwa and Nigig brought giizhikaatig to the Anishinaabeg to help them, and taught them many uses of giizhikaatig, including how to make medicines out of this tree.

The Anishinaabeg use giizhikaatig as an important healing ingredient in a variety of medicines, including one that removes warts and one that heals chapped skin. Both of these medicines are unguarded gikendaasowin, and Keewaydinoquay gives these recipes in the video *Native American Philosophy and Relationships to Plant Life* (M. Geniusz, pers. comm.; Keewaydinoquay 1986). The wart medicine combines ground cedar foliage with apple cider vinegar. This medicine is applied directly to the wart and tied with a bandage. This was one of the first mashkikiwan that I was taught how to make because my mother was using it to treat a plantar wart on my foot. It removed the wart without the pain caused by physician-prescribed wart medicines (M. Geniusz 2005, 52).[23] Another medicine made from giizhikaatig Keewaydinoquay calls "Cedar Lemon Balm."[24] This name comes from the two main ingredients of this medicine: cedar and lemons. Before the Anishinaabeg had access to lemons, they used common wood sorrel as the acid for this medicine.[25] Mary Geniusz says that Keewaydinoquay told her she thought that Nodjimahkwe, who was skilled at modernizing recipes, was the one who changed this ingredient from wood sorrel to lemons, which were readily available by the time Keewaydinoquay first learned this recipe. The switch to lemons makes sense, Geniusz adds, because, although wood sorrel is a very

23. See the appendix for a "Cedar Wart Medicine" recipe.

24. See the appendix for a "Cedar Lemon Balm" recipe.

25. Geniusz identifies two species of wood sorrel, *Oxalis montana* Raf. and *Oxalis stricta* L., as the possible ones used in the "Cedar Lemon Balm" recipe (M. Geniusz 2005, 52). Geniusz says, however, that in her class notes she wrote that when she asked Keewaydinoquay what was used in this medicine before the Anishinaabeg had access to lemons, Keewaydinoquay said common wood sorrel, which she identified only by the genus name, *Oxalis* (M. Geniusz, pers. comm.). Both of these species live in the Great Lakes region, and they are both native to North America. They are both said to have a sour taste, and they look very similar (Niering and Olmstead [1979] 2001, 670–71; USDA, NRCS 2006).

common plant, it can only be used in this recipe when it is green. This limits the times when one can make this mashkiki, and because it does not keep for more than one month, it is not always available if made with wood sorrel. Lemons, however, are currently available to us twelve months of the year, and so it is much easier to grind up lemons for this recipe than to use wood sorrel (M. Geniusz, pers. comm.).

Geniusz writes that Keewaydinoquay often taught her university students how to make this medicine, and she encouraged them to use it as a general skin care product. "Cedar Lemon Balm" is made from equal parts ground cedar foliage and lemon. Once ground, these ingredients are mixed with just enough grease so that the mixture can easily spread on the skin (M. Geniusz 2005, 51–52). Keewaydinoquay says that bear grease used to be used in this recipe before the Anishinaabeg had alternatives that would work just as well (1986). Keewaydinoquay told Geniusz that when Nodjimahkwe taught her to make this medicine, she used bear grease, but that Keewaydinoquay changed the recipe to Crisco so that she could make this mashkiki without affecting the dwindling bear population. Keewaydinoquay's mother used to use porcupine grease instead of bear grease in this recipe because it does not have any odor or taste as does the bear grease. Geniusz adds that any edible grease can be used in this recipe (2005, 51–52; M. Geniusz, pers. comm.).

In her university courses, Keewaydinoquay often added other ingredients, including olive oil and glycerin, to this skin cream. Geniusz explains that the only essential ingredients of cedar lemon balm are the cedar foliage and the acid because these two ingredients soften the skin. The grease is used to hold this mixture against the skin. Geniusz adds to this information that in a survival situation, for example if someone is walking a long distance and his or her feet begin to crack and bleed from the walking, smashed cedar foliage alone, applied to the skin, can heal it much faster than applying nothing at all. Long ago, an Anishinaabe traveling with cut feet would smash up cedar foliage with a rock and tie it onto the feet with a thin strip of birch bark, or by some other means, and then continue the journey. Walking on the cedar in that way would keep the healing oils of the cedar against skin, and even more of the oils would be released by the pressure of walking (M. Geniusz, pers. comm.).

## NINDINAA NOOKOMIS, NINDABANDENDAM: WE SAY NOOKOMIS, WE SHOW RESPECT NOOKOMIS SA GIIZHIK, GICHITWAAWENDAAGOZIWIN

The final Ojibwe line of "The Cedar Song" before the refrain is "nindinaa nookomis, nindabandendam," corresponding to the English line "we say nookomis, we show respect." Literally this Ojibwe line means "I say to her, nookomis, I am respectful." The refrain, "nookomis sa giizhik, gichitwaawendaagoziwin," as explained in chapter 2, has a similar meaning. The refrain emphasizes that this tree is our grandmother, and this is holy. Of giizhikaatig, Keewaydinoquay says, "It really does not matter what we call her as long as we treat her with respect" (1986). She says that names showing a familial relationship are often added to the Ojibwe names for cedar and birch because these trees are so highly respected in anishinaabe-inaadiziwin (1988). She calls cedar "Nookomis giizhik, Grandmother cedar," and birch "Nimishoomis wiigwaas, Grandfather birch" (1988).

I have never heard any other Anishinaabeg refer to either of these trees as Nookomis or Nimishoomis. Keewaydinoquay says that these terms of respect were once used frequently by the Anishinaabeg, but that during the lumbering days they stopped being used because it was difficult to think of these trees as "my grandmother" or "my grandfather" when cutting them down to make a living (1989a). The teaching of respect for these trees certainly exists in inaadiziwin, as does the dibaajimowin that all life should be respected. Examples of these teachings come from elders as well as from written texts. As explained in chapter 2, the protocols of izhitwaawin teach us to respect other beings and accept that, whether we call them by a familial name or by a more general name, these other beings can assist us in our physical and spiritual lives because we are all interconnected. As will be explained in the conclusion, we need to look at differences, such as this one, between inaadiziwin and non-native worldviews to find our greatest weapons against further colonization.

# Conclusion

If cultural revitalization is our goal, then it makes sense that all of this information, both the very detailed and the not so detailed, must be brought back into the context of izhitwaawin. In order to do that, we must understand how botanical knowledge is viewed through inaadiziwin and maintained within izhitwaawin. I began this process in chapter 2. If we understand at least some of the basic principles of inaadiziwin—the knowledge-keeping system out of which this information came—then we will be in a position to assess the value of this information for cultural revitalization programs. If we do not know, for example, the proper protocols for gathering botanical materials within izhitwaawin, then knowing how to use those materials can only help our cultural revitalization to a certain degree. Being able to make a mat out of the bark of cedar is an important part of izhitwaawin, but if we do not know how to properly address the cedar to ask for her physical and spiritual assistance, then we will be missing a key component of inaadiziwin. Without this vital information, our cultural programs are only craft workshops or art classes where participants learn skills without understanding the cultural context and the worldview out of which those skills come. Our people will be able to understand something about izhitwaawin, and having even this small piece of gikendaasowin may help them, but they will not be able to utilize the full healing abilities of izhitwaawin. They will not be able to completely undo the colonization process within themselves.

Vine Deloria, Jr., argues that the strength of tribal knowledge lies not in how it is similar to "science" but where it is different. He writes, "We may grant that the energy described by quantum physics appears to be identical to the mysterious power that almost all tribes accepted as the primary constituent of the universe. But what does this conclusion say about theories of disease, powers of spiritual leaders, or interspecies communications with sympathetic birds and animals?" (2001a, 5). I began to answer some of these questions in chapter 2. Deloria sees these differences as one area in which tribal knowledge can aid the rest of the world because tribal knowledge is willing to take into account many phenomena that science ignores. He argues that Indian students can "bring to science a great variety of insights about the world derived from their own tribal backgrounds and traditions" (2001b, 28). He uses these questions of differences between native and non-native philosophies to encourage Indian scholars to bring their tribal worldviews to their scientific investigation so that they may better accomplish the goals of those investigations and so they may help the world "make more constructive choices in the use of existing Indian physical and human resources" (2001b, 23–28). Deloria's suggestions are also applicable to the topic of decolonization. In relation to the decolonization of botanical gikendaasowin, these differences between scientific and tribal knowledge are an area in which we can find the greatest strength of inaadiziwin. This piece of it is so powerful that it not only helps us to decolonize botanical gikendaasowin, it can also help us to decolonize ourselves.

In Ojibwe one describes another person as "colonized" by saying, "*Zhaaganashiiyaadizi*," which means that person tries to live his or her life as a non-native, at the expense of being an Anishinaabe. Such a person is living a life of unbalance. One of the major goals of Biskaabiiyang research is for an individual to bring him or herself back to inaadiziwin. Biskaabiiyang research attempts to correct the state of *Gaawiin wiiskiwizisii wii-maazhised,* which means a person is not living in balance; his or her behavior is unnatural and therefore could lead to death ("Anishinaabe Wordlist" 2003). Reaching a state of *wiiskiwiziwin* (balance) is an important part of the decolonizing process. Keewaydinoquay ended every one of her ceremonies by having all the participants join hands and say, "Blessings and balance; Balance and blessings; for from balance comes all blessings. Ahaw." This dibaajimowin exemplifies one of the major differences between inaadiziwin and non-native philosophies: the idea that balance is the source of blessings. From balance comes good. From balance comes health. Keewaydinoquay often said that the reason people are not always cured by non-native medicine was that they were healed physically but not spiritually (1991a). The great virtue of plants is that they can, when asked properly, help us maintain our own balance by bringing us both physical and spiritual healing. We maintain this state of balance when both our bodies and our spirits are in good health. There are also certain plants that specifically help us to restore balance in our lives. As stated in chapter 4, giizhikaatig is our line of communication with the manidoog and the aadizookaanag, those beings who can help us when we need assistance in bringing balance back to our lives. One way to do this is to make an offering to this tree and ask through her for the required assistance. Other plants and trees can assist us as well. Ken Johnson, Sr., says that ***mina'ig***,[1] which he identifies as the highland spruce, has the ability to help us when we have a great sadness or mental anguish. He says that to ask mina'ig for this help, one goes up to this tree, lays down asemaa, and then explains the problem to this tree (Johnson 2006).

Colonization has forced many Anishinaabeg, and other indigenous people, into a state of unbalance. Drug and alcohol abuse are prevalent in many anishinaabe families. So are feelings of hopelessness, worthlessness, and thoughts of suicide. Some parents are so lost in their own inner turmoil that they physically cannot raise their children. Children grow up not only knowing nothing about what it means to be Anishinaabe, but also not knowing what it means to have parents. For a lot of Anishinaabe children hunger is a fact of life, not just something seen in infomercials requesting funds for children on other continents. I have been a part of several programs revitalizing izhitwaawin. I know that teaching an Anishinaabe child or adult how to make a pair of moccasins or how to speak a few words of Ojibwe can be a powerful thing. Many of the children and adults with whom I have worked had no pride at all in being Anishinaabe, but learning these things seemed to give them some sense of cultural pride. I am not sure how long term the effects of these feelings were, however, especially knowing that shortly after the conclusion of our programs, many of the adults involved were back in the bars. Our programs had some effect, but they did not decolonize our participants.

I started this research hoping to create resources that could be used for programs revitalizing izhitwaawin. I wanted to decolonize texts containing

1. Rose identifies mina'ig as the white spruce, *Picea glauca* (Moench) Voss (pers. comm.).

botanical gikendaasowin because I saw my fellow Anishinaabeg, especially those involved in revitalization programs, using some of the colonized texts described in this research to learn about how our people use plants and trees. I wanted to create a resource for these individuals and programs to replace these colonized texts. The first step in researching or using this knowledge, either information from a colonized or a decolonized text or from an Anishinaabe elder, is to decolonize oneself. I can write about the gifts that many plants have offered the Anishinaabeg, but if the reader looks at it through the eyes of the colonizers, then the information I am sharing with him or her will not fully help that person. Through colonization we have been trained to refer to anything that is not human as "it." Although in English "it" does not necessarily mean inanimate, the use of this pronoun places the entire nonhuman world in a separate category, generally viewed as being beneath us. Following this line of thinking, plants, trees, animals, and the rest of Creation were placed here for us to use. As Deloria argues, not opening our eyes to other worldviews greatly limits our creativity and abilities to solve problems (Deloria and Wildcat 2001). From the perspective of izhitwaawin, if we cannot recognize that plants and trees have the ability to help us reclaim our balance, hope, and sense of self-worth, then our revitalization programs will never be fully effective. So ultimately, we need to decolonize gikendaasowin as it is preserved within the academic record and as it is preserved within ourselves.

Colonization has been happening to us for many generations; decolonization will not happen quickly or easily. We need to be patient and accept that, although we cannot expect to be completely decolonized all at once, every step we make in that direction is something of which to be proud. One of these steps is seeing plants and trees through the perspective of inaadiziwin. We are surrounded daily by plants and trees, beings who can help us. By acknowledging and accepting this help we will have begun the journey back to our teachings and ourselves. We will bring our people and ourselves that much closer to becoming decolonized.

# Appendix | Glossary | References | Index

# Appendix

## *Instructions for Working with Giizhikaatig and Wiigwaasi-mitig*

Instructions are given on the following pages for working with giizhikaatig and wiigwaasi-mitig to make some of the medicines and other items mentioned in chapter 4. These instructions are given so that readers will know how to make these things, rather than just knowing that they can be made.

According to the protocols of izhitwaawin, one must make an offering of asemaa or kinnikinnick and explain to the tree or plant why it is necessary for that being to be used for any purpose. This offering and explanation should be made before following any of these recipes or instructions. Also, before working with any botanical material, it is important to be aware of pollutants that may have affected the plant or tree one intends to use. Finally, although the recipes on the following pages are relatively safe, individuals can have allergies to anything, so do not use any of these recipes, especially the medicines, in large quantities until it is certain the person using them is not allergic to the materials involved.

### DIRECTIONS FOR MAKING A TEA FROM GIIZHIKAATIG

As mentioned in chapter 4, the Anishinaabeg make a tea from giizhikaatig, which is high in vitamin C and can be used to treat colds, alleviate stomach gas, and take the "fishy" taste out of one's mouth after a fish dinner. To make tea from giizhikaatig: boil some water, let the water stand for five minutes, and then pour it over some cedar foliage. The water needs to stand for five minutes before being poured over the foliage so that more vitamin C is released into the tea (M. Geniusz, pers. comm.). Keewaydinoquay cautions her students not to leave the foliage in the warm water longer than ten minutes or the turpenes, which are present in giizhikaatig and other conifers, will be released into the hot water along with the vitamin C. She explains that these turpenes are not a healthy thing to drink, as they are the source of turpentine (1988). I usually leave the foliage in my tea for about five minutes. Mary Geniusz warns that a pregnant woman should not drink a strong tea of cedar or it may cause her to lose the child she carries (2005, 47).

### RECIPE FOR "CEDAR WART MEDICINE," A MEDICINE FROM KEEWAYDINOQUAY

Giizhikaatig is also an important ingredient in a medicine used to remove warts. To make this medicine, one separates the foliage of giizhikaatig from the woody twigs and grinds it. Then one mixes the ground foliage with apple cider vinegar. Geniusz writes that these two ingredients should be mixed together "until the pulp is mushy and sticky" (2005, 52). Keewaydinoquay describes this mixture as "sort of a saucy mix" (1986). Basically, the resulting mixture is sticky and gooey.

To apply this medicine, one puts a poultice, or glob, of this mixture onto the wart and covers it with a bandage. Geniusz writes that this should be done at night (2005, 52). When I had a wart on my foot, I reapplied this mixture every time I washed my feet, so it was always on my foot. This medicine keeps for a couple of weeks when stored in a glass jar in the refrigerator. A relatively new wart, when treated this way, should come off within a week or two. If it is an older wart, it may have to be pulled out with the fingers once it has been loosened by this medicine. The wart on my foot had not been there very long, and I was able to pull it out within three weeks, and it never reappeared. This medicine was not painful at all to apply to the wart, as is the medicine that doctors prescribe for removing warts. When I tried a physician-prescribed medicine on another wart, it ate away so much of the surrounding skin that I had to stop using it, and when that wart finally went away, it took a week for my skin to heal from the effects of the medicine.

If the wart is very old and will not come loose once this medicine is applied, then one applies sundew (*Drosera intermedia* Hayne)[1] to the root after removing the top of the wart with the "Cedar Wart Medicine" (Keewaydinoquay 1986; M. Geniusz 2005, 52–53). Keewaydinoquay describes sundew as "a swamp plant that is actually a meat eater" and says that when applied to the skin, the sundew will eat

1. Sundew (*Drosera intermedia* Hayne): Geniusz is not sure if she made this identification or if Keewaydinoquay did (pers. comm.). Meeker and colleagues list another species, *Drosera rotundifolia,* as being used by the Anishinaabeg, but they do not say how this plant is used (1993a, 192).

away the old wart, making room for new tissue to grow in its place (1986). Geniusz showed me this plant once when we were in a sphagnum swamp in northern Wisconsin. She writes that this plant should be gathered fresh, and the little droplets, which look like dew, found on the tiny hairs of this plant should be squeezed onto the wart. These little droplets attack the root of the wart so that it will come out (2005, 53). It should be noted, however, that *Drosera intermedia* Hayne is endangered in many areas, and one should check the U.S. Department of Agriculture Web site before picking this plant in his or her area.[2] In izhitwaawin, this is a secondary cure, tried only if the first cure, the wart medicine made from cedar, is ineffective. In today's world we have so many other ways of treating warts that we have alternatives to uprooting this endangered plant. Killing the last of one kind of plant in an area, as described previously, goes against the protocols dictated by izhitwaawin. Also, one should know that this plant is not something to be handled without caution. Keewaydinoquay warned her students that this plant can dissolve skin, and if one is administering this plant to a patient, he or she should be careful not to get these little droplets on his or her skin as well. If an unaffected piece of skin does come in contact with these droplets, it should be washed immediately (M. Geniusz, pers. comm.).

### RECIPE FOR "CEDAR LEMON BALM," A MEDICINE FROM KEEWAYDINOQUAY

Giizhikaatig is an important ingredient in "Cedar Lemon Balm," a mashkiki used to heal chapped skin. To make it: first grind, grate, or chop equal parts cedar foliage and lemon. Use the entire lemon except for the seeds. Electric blenders work nicely for this process, but I have also used mechanical meat grinders and electric coffee grinders.[3] To tell if the mixture contains the right amount of cedar and the right amount of lemon, taste it. If it tastes like both lemon and cedar, then the mixture has the right amount of both ingredients. If one taste dominates the mixture, add more of the other ingredient until the two can be tasted equally. Then mix just enough Crisco into the mixture so that it can be easily spread on one's skin, without large clumps of the cedar and lemon mixture falling off right away. When making a large

2. To check if a particular plant is endangered, see the U.S. Department of Agriculture Web site: http://plants.usda.gov.

3. I recommend a coffee grinder that is only used for grinding herbs; otherwise coffee will get into the skin cream.

quantity of this medicine, such as with a class of about thirty people, a sixteen-ounce Crisco container provides enough grease to use with three ground lemons and a quarter of a grocery store bag of cedar foliage.

When Mary Geniusz was working as Keewaydinoquay's teaching assistant at the University of Wisconsin-Milwaukee, she used to distribute handouts of this recipe to the students. In the recipe that Keewaydinoquay gave to her students, she included other ingredients that are not essential to this medicine, but that are readily available in today's world and are helpful to the healing process. This recipe includes adding a little olive oil, approximately one teaspoon, which Geniusz explains has healing properties in it that help make this medicine even more effective. The olive oil is helpful if the cedar being used is not very oily, especially during the winter. This recipe calls for ten drops of benzoin, which acts as a preservative. Geniusz says, however, that it only preserves the lotion for a short time and is not essential to the final product. Keewaydinoquay's recipe also includes glycerin, which also helps to soften hands. Keewaydinoquay used to send Geniusz to buy the glycerin for the ethnobotany labs, specifying that Geniusz was to buy only vegetable-based glycerin rather than the cheaper kind, which was animal-based. Keewaydinoquay liked to use glycerin in many of her recipes, but some of her students did not want to include glycerin in "Cedar Lemon Balm" because it can dry the skin (M. Geniusz, pers. comm.).

Geniusz explains that the only essential ingredients of "Cedar Lemon Balm" are the cedar foliage and the acid from the lemons because these two ingredients soften the skin. Even the grease is not an essential ingredient, as it is present primarily to hold this mixture against the skin. I asked Geniusz if the medicine would work better if it was not diluted by being mixed with the grease, and she said that it does not matter if it would work better without the grease because the medicine will only work if the patient uses it. Keeping a ground mixture of cedar and lemon against the skin without grease is difficult, but applying cream is not. If the person using this medicine will use it more often because it is easy to apply, then the grease is definitely needed and the medicine will work better for that person because it is mixed with grease (M. Geniusz, pers. comm.).

I have made "Cedar Lemon Balm" several times using just giizhikaandag, lemons, and Crisco. I have applied this mixture to my hands before going to bed when they were chapped so badly that they were bleeding, and in the morning they were nearly healed. Geniusz suggests that one apply "Cedar Lemon Balm" to the feet or hands before going to bed, and cover these areas with socks so that the little pieces of cedar foliage will not flake out onto the sheets during the night (2005, 52). Cotton gloves, such as those available in many cosmetic departments, should work just as well. I have also applied "Cedar Lemon Balm" to my skin and then covered my skin with clothing, which successfully held the mixture in place.

To preserve it for approximately two weeks, keep this mashkiki in a glass jar in the refrigerator. Keewaydinoquay insisted that all her mashkiki be kept in glass jars because she worried that the chemicals in her mashkiki would react with plastic, causing complications (M. Geniusz, pers. comm.).

### DIRECTIONS FOR GATHERING THE INNER BARK OF GIIZHIKAATIG

The inner bark, also called the cambium,[4] of giizhikaatig can be used as an emergency food and to make mats and rope. Bark can be gathered from a living or dead giizhikaatig. In her university lectures, and in the video *Native American Philosophy and Relationships to Plant Life*, Keewaydinoquay shows slides of how one peels cedar bark from a downed tree. In these slides, the person peeling the bark uses an ax to pry the bark off the tree bit by bit in sheets. Keewaydinoquay says in this video that bark is especially easy to peel from a tree that has been lying in a lake or in a swamp for a while (Keewaydinoquay 1986; M. Geniusz, pers. comm.). I have peeled bark from a tree that had fallen months earlier, and only the portion that was lying in the mud for an entire winter was possible to remove. The rest of the bark was very dry and could not be pried loose. From the portion I could peel, I was only able to get small strips off at one time, but I was only using a pocketknife, not an ax.

When peeling bark from a living giizhikaatig, one should keep in mind that stripping all of the bark off a living tree will kill the tree, and within the context of izhitwaawin, one does not take a life without justification. Volney Jones writes that Neil Boyer, of the Garden River Reserve in Ontario, told him that a cedar tree from which bark was gathered would continue to live if one-third of that tree circumference was still covered in bark (1946, 343). Mary Geniusz says that when taking bark off any tree it is important to be careful not to girdle the tree; that is, not to cut the inner bark off in a ring around the tree. If one is not going to use the entire tree, following Boyer's advice, only stripping two-thirds of the bark would be a way to avoid taking the life of giizhikaatig unnecessarily. All the written sources stress the

4. The inner bark and the cambium are the same thing. Crockett describes the cambium as "only one cell thick" and says it is under the phloem, which is under the outer bark, the part of the tree that we can see without peeling away any bark (1972, 20). Lee Allen Peterson describes "the inner bark (cambium)" of trees being used as flour (1977, 295). Keewaydinoquay uses the words "inner cambium" when describing this emergency food source (1988).

importance of removing the bark from giizhikaatig while the sap is flowing in the tree. Depending on the area in which one is peeling, this can be as early as May and as late as August (V. Jones 1946, 341, 343; Petersen 1963, 221). I have peeled cedar bark in northern Wisconsin during the month of June, the same time when the bark of wiigwaasi-mitig is ready to peel, and during that time the bark slipped easily from the tree; it did not even get stuck around the tree branches. When I peeled this bark, both the underside of the bark and the exposed wood of the tree beneath were wet. In this condition bark is easy to bend into a spiral for storage. The bark I was peeling was from a tree that we had chopped down to make a canoe, so I just started peeling this bark from where the tree had been severed from its roots with the ax. Volney Jones describes how Boyer gathered bark from a living tree: "With an ax he cut through the bark about two feet above the ground level, notching the bark transversely for approximately two thirds of the distance around the tree. He then pried the bark free, so that it could be grasped with the hands and pulled loose from the tree in long narrow strips as far as the first limbs, where it broke off" (1946, 343). Jones says that the trees Boyer peeled were "six inches to one foot in diameter and that they had no lower limbs" (1946, 342–43). I was peeling a much smaller tree, which had lots of tiny branches coming off of its trunk. I clipped off the small branches, so the bark only had to peel over the stubs where the branches had been, not over the branches themselves. At these stubs, the bark just formed a hole and continued to peel. Petersen says that Mrs. Goodsky and her assistants cut off all the branches they could reach and then made a one-and-one-half-inch-wide horizontal cut approximately sixteen to twenty-four inches from the ground. They loosened the bark with an ax blade and used their fingers to pull off "a short vertical strip," and they continued to gather bark from either side of this exposed section of the tree (1963, 221–22).

Once peeled off the tree, the outer and inner barks need to be separated. Only the inner bark, or cambium, of giizhikaatig is used as an emergency food and as a material to make cordage and mats. Jones and Petersen both describe the people they worked with separating the outer and inner cedar bark soon after gathering it.[5]

They also both describe a similar process of dividing the two barks by bending the strips of bark so that the outer bark broke into pieces, which were then peeled

5. Jones describes the two layers being separated before the strips of bark are rolled up to be transported, and Petersen describes the two layers being separated after the bark has already been rolled up and transported to another location (V. Jones 1946, 344; Petersen 1963, 222–24).

from the inner bark (V. Jones 1946, 344; Petersen 1963, 222–24). I have separated the inner and outer barks long after they were gathered by soaking the bark overnight so that it was pliable. Mrs. Bellanger, who advised V. Jones, says that dried cedar bark can be soaked to make it pliable once more, but when this is done "the bark assumes an undesirable red color" (V. Jones 1946, 345). It should be noted, however, that Bellanger gathered and dyed her strips for weaving a cedar mat, not for use in a survival situation.

### DIRECTIONS FOR MAKING ROPE OUT OF GIIZHIKAATIG

Rope can be made out of the inner bark of giizhikaatig. This rope can be used for a variety of things, including making a snare or a bowstring. As with any item made from a botanical fiber, there are different ways to work with cedar bark to make a cord or rope. Several of Keewaydinoquay's oshkaabewisag taught me how to make cedar rope when I was twelve, and they said that they learned how to make this cedar rope from Keewaydinoquay. First one rolls the inner bark against one's leg, so that it forms a round piece of cord, which is then folded in half. Each end of the cord, starting at the fold, is then twisted individually in opposite directions, and then the two halves are twisted together. For example, twist the right half clockwise and then the left half counterclockwise, and then twist the two halves together clockwise. Additional fibers are twisted into the cordage as needed.

### DIRECTIONS FOR MAKING MATS OUT OF GIIZHIKAATIG

Mats made of giizhikaatig are often used to cover the lower portion and the floor of the waaginogaan. There are many ways to weave these mats, and what follows are descriptions by two researchers, Volney Jones and Karen Daniels Petersen, on two very similar methods of weaving mats out of giizhikaatig. Following this section there are diagrams, drawn by Annmarie Geniusz, of the weaving process. These are based mainly on the descriptions given by Jones and Petersen. Once the light-colored inner bark of giizhikaatig is gathered and processed, the pieces of bark are divided into strips of uniform width and length that will be used to weave the mat. V. Jones describes Mrs. Bellanger and her daughter cutting four-foot-long strips for the warp and cutting longer, six-foot-long strips for the weft (1946, 345).[6] Once the strips are

6. Petersen also describes two lengths of strips being cut for the weaving (1963, 224–25).

cut, the weaver can dye them various colors. When preparing the strips for weaving, Mrs. Bellanger measures the weft strips, which will be woven through the warp, so that they are all equal lengths. She prepares the warp strips--the vertical strips on which the weaving is done--by thinning four inches off one end of each strip. She does this by cutting into the bark with a knife about halfway through its thickness, and then folding the bark where it was cut. This step causes the bark to break so one can strip off the top layer. The other end of each strip is "tapered to a point" (1946, 345). Mrs. Bellanger hangs the warp of her cedar mat on a basswood twine.

Before being put on the frame, the warp of the cedar mat is created by attaching each strip of the warp to the basswood cord. First, the basswood twine is tied to an object that will hold it steady while the weaver makes the warp. According to Jones, Mrs. Bellanger fastens the end of the basswood twine to a wall with an ice pick. She leaves four feet of basswood twine dangling from where it is tied to the ice pick, and the other end, which is sixteen feet long, she uses to tie the strips of cedar bark to form the warp. She folds the thinned portion of the warp strips around the basswood twine about four inches from the end. The shorter end of the strip hangs on the back, and the longer end hangs on the front. Mrs. Bellanger brings the shorter end under the twine, across the front piece, and lays it against the basswood twine. Then she folds the next two cedar strips over the end of this strip that lies against the basswood twine. As the mat progresses, the short end of each strip is bound by the two beside it. While she is doing this step, V. Jones reports, Mrs. Bellanger keeps a supply of cedar warp strips around her shoulders so that she can easily grab them to tie the next piece on. She also loops the finished portion of the warp around the ice pick as it becomes too heavy under the weight of all the warp strips added to it. When she has to interrupt this work, she ties the loose ends of the warp with a piece of basswood fiber, and lets the entire warp hang from the ice pick. She takes this fiber off when she continues her work. When she is done making the warp, Mrs. Bellanger ties the loose ends with basswood twine, unfastens the warp from the ice pick, and ties it on the weaving frame. This frame consists of two uprights with one crosspiece over the top, and V. Jones notes that the frame used to weave the cedar mats is the same kind of frame he saw used to weave rush mats (1946, 345–50).[7]

7. The method that Jones describes is, of course, only one way of weaving a cedar bark mat. Petersen, for example, describes Mrs. Goodsky using commercial twine instead of basswood twine to tie the warp strips of her mat, and Mrs. Strong, another weaver, used a strip of cloth instead of any kind of twine as the foundation for the first selvedge (edge) of the mat (1963, 226). Mrs. Goodsky, according to Petersen, tied her warp strings to the twine

Once the selvedge for the warp is completed, the warp is lashed to a frame on which the actual weaving will take place. V. Jones describes Mrs. Bellanger's mat as being tied onto the crosspiece of a frame that was constructed before she began making the warp (1946, 345–46).[8] The basswood twine that will form the left side of the mat reaches all the way from the top of the warp to the ground. The one that will form the right side of the mat, however, reaches farther, as it will form both the right side of the mat and the bottom of the mat (1946, 346).[9] Petersen says that the warp is lashed to the crosspiece with a piece of string, which is wrapped around both the top twine of the warp and the crosspiece approximately every two to three strips (1963, 228–29).

When weaving the cedar bark mat, simply weave the weft strips over and under the warp strips. The end of the weft strip is fastened to the basswood twine hanging on the left side, just as the warp strips were fastened to the basswood twine on the top of the weaving (V. Jones 1946, 350). The weft strip is woven over one warp strip and then under the next and then over the next, and so on, until the end of the row is reached. At the end of the row, the weft strip is fastened to the basswood twine hanging on the right side. The woven weft strip is fastened to the right side with the same procedure as it was fastened to the left side, but Petersen notes that the direction is reversed, and that the weft is also trimmed to about three inches before it is fastened to the twine on the right side. The next row is done the same way, except it is alternated; where the first row went over one of the warp strips, the second row goes under that same warp strip (V. Jones 1946, 350–51; Petersen 1963, 228–29). There are various designs that can be added to the mat by means of strips dyed in

---

following the same method that Mrs. Bellanger used, but Mrs. Goodsky fastened her twine to a chair instead of a wall, and she worked from right to left folding her strips over the twine, instead of working from left to right as Mrs. Bellanger did. Mrs. Goodsky also had someone move the chair onto which her warp was fastened, rather than looping the warp over an ice pick as Mrs. Bellanger did (Petersen 1963, 226–28).

8. Mrs. Goodsky's mat is tied onto the crosspiece of the frame before that crosspiece is lashed to the uprights of the frame. Petersen suggests that this movable crosspiece is easier to tie the warp onto because it allows the weaver to tie the warp on at any comfortable height (1963, 228–29).

9. Petersen describes the ends of the twine on which the strips of the warp are tied as being tied in a bow near the ends of the crosspiece of the weaving frame. During the weaving, these ends of the twine are used to tie the ends of weft, to finish off the ends of the mat. Another piece of regular string is then used to tie the warp to the crosspiece (1963, 228).

different colors and by means of changing the way the weaving is structured from one row to the next. For example, Densmore describes different patterns being created by passing one weft strip over two or more warp strips instead of just over one ([1929] 1979, 156–57).

When the mat is woven to its desired length, only the bottom of the mat has to be finished off, as the top of the mat was finished before the mat was put on the frame, and the sides of the mat were finished as the weft strips were woven. V. Jones warns that the mat must be finished off before the weaving reaches farther than four inches from the bottom of the warp strips because this much material is needed to finish off the weaving (1946, 352). V. Jones describes one nail being driven into each of the uprights directly across from the last row of weaving. Then the cord on the right side of the weaving is looped around the nail in the right upright and brought across the warp where it is tied on the nail in the left upright. This is the same cord that the wefts were attached to on the right side and the same cord that was purposely made longer than the cord on the left side (1946, 352–53).[10] It is this cord to which the bottom edge of the mat will be tied. The bottom edge of the mat is finished off just as all the other edges were, with one strip being folded over the twine and then brought around itself, and laid on the twine, and the next strip going over this twine (V. Jones 1946, 353; Petersen 1963, 231). At the very edge of the mat, once the entire bottom is tied off, Mrs. Bellanger braids the end of the last warp piece, the last one to be fastened to the bottom cord, with the end of the lowest piece of the weft, which was split so that it could be braided. She then turns this braid to the back of the mat and fastens it there with a piece of basswood cord. When this mat is put on the floor, this braid will be on the bottom of the mat and out of sight. V. Jones also says that the loose cord on the left of the mat, presumably the one that was brought across from the right to form the cord on which the bottom of the weaving was fastened, is looped around the border of the mat a couple of times before it is tied and cut off (1946, 353).[11]

Both Jones and Petersen offer some good advice about how to fix inevitable problems that arise while weaving. Jones warns that if only one person is doing the

10. Petersen says that Mrs. Goodsky does not use nails but simply brings the cord from the right side to the left side of the weaving frame (1963, 231).

11. Petersen says that Mrs. Goodsky, upon finishing off the bottom of her weaving, ties all the ends "together firmly with several turns of sewing thread and several knots" and then cuts them so that they are one-half inch long (1963, 231).

weaving, the weft pieces on either side must be tied in place because one cannot hold them while weaving the middle. On the mat he was weaving this step was not necessary, as there were three people weaving: one to hold onto fastened pieces on each side and one to do the weaving in the middle (1946, 351).[12] When leaving the weaving for a time, Mrs. Goodsky prevents the weft from sagging by making a double bend in a warp string, about two inches from the bottom of the last weft row, and then pushing this bent piece up between the last two weft strips. She repeats this process across the entire weaving about every three to six inches (Petersen 1963, 230). As with any weaving project, keeping the margins of the cedar mat straight is important. V. Jones describes a thong being tied around the edge of the mat and the upright beside it to keep the tension even, and this thong is moved down as the weaving moves down (1946, 351). Petersen says that Mrs. Goodsky ties the twine on which the edges of the weft are fastened tautly and keeps them tied no more than twelve inches below the portion she is weaving. Petersen says this step may help Mrs. Goodsky keep her margins straight. When a weft breaks, a few inches of the weaving are ripped back and another strip of cedar bark is inserted. Then the two are used together as one doubled strip for a few inches (1963, 230).

Another important thing to remember when weaving a cedar bark mat is that the materials being used in the weaving must remain wet or they will become less flexible and crack. The materials must also be wet for making rope out of cedar bark. For this reason, cedar mats are often woven in the shade to prevent the materials from drying out too fast (V. Jones 1946, 345–46; Densmore [1929] 1979, 157). Densmore provides a photograph of a structure built for mat weaving, and this structure would obviously provide weavers with a shady and sheltered place in which to do their weaving ([1929] 1979, plate 62). Petersen says that Mrs. Goodsky, who did her weaving inside her house, moved the frame of her weaving into the shade when the sun shone in on it. Petersen also describes methods of keeping the weaving materials damp. The weft strips waiting to be used are kept coiled in a pan with water and sometimes patted with a damp cloth. The parts of the warp that have not yet been woven are kept damp by throwing water from a cup on them (1963, 229).

12. Petersen says that Mrs. Goodsky does not tie the end of her wefts and they do not come unfastened. Petersen suggests that the difference between the mat-weaving techniques Jones describes and those used by Mrs. Goodsky might be because Mrs. Goodsky's "more pliable and narrower strips (about half as wide) could be pulled more tightly at the selvedge" (Petersen 1963, 230).

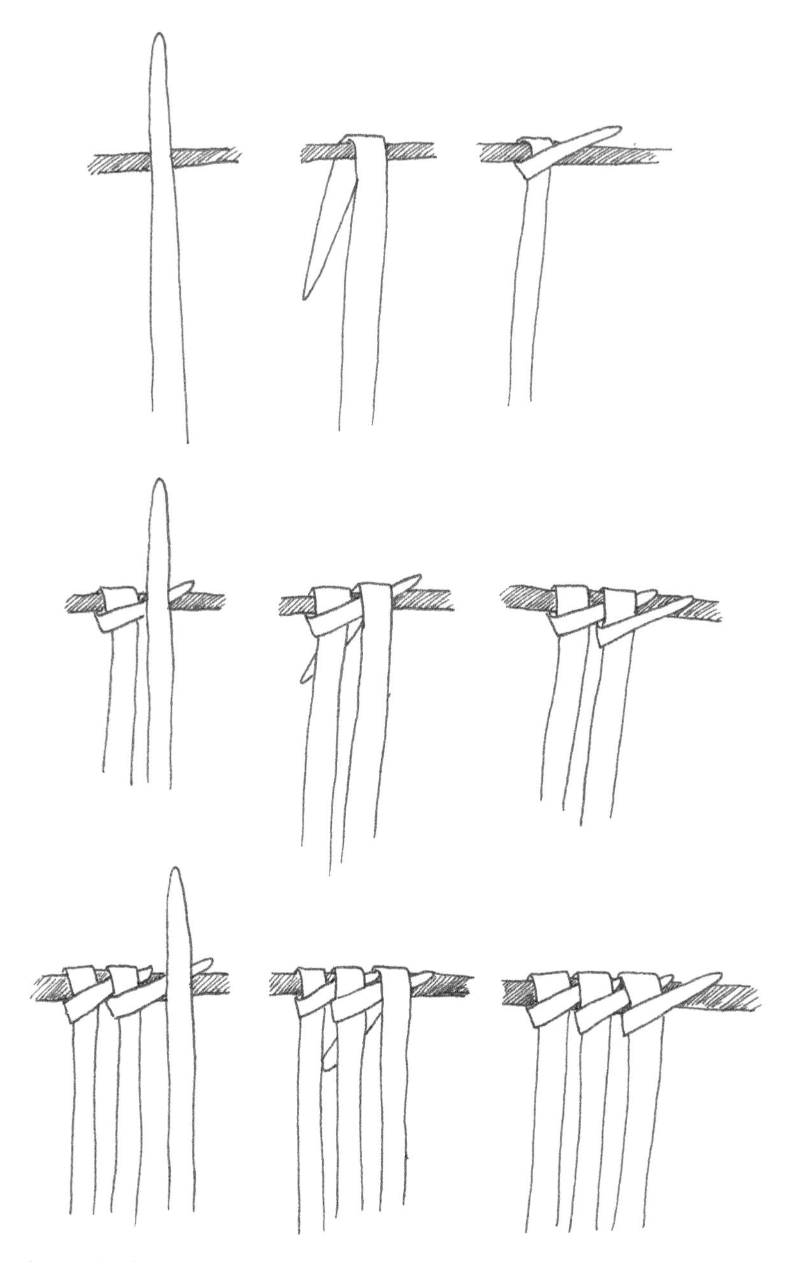

5. Detail of attaching warp strips to basswood cord. © Copyright Annmarie Geniusz. Used with permission.

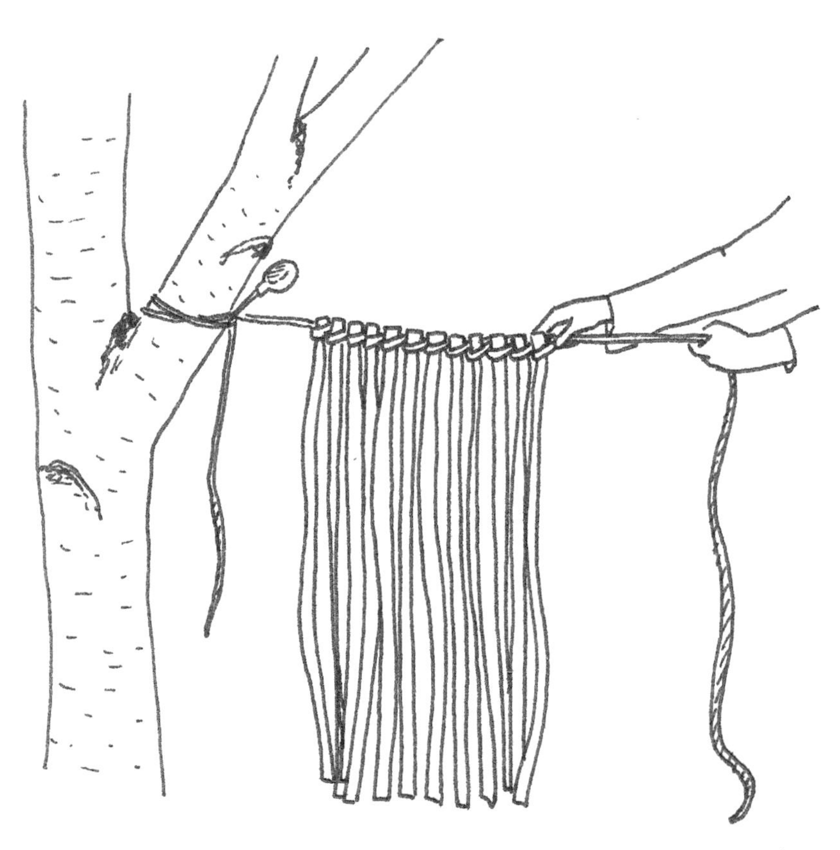

6. Attaching warp to basswood cord. © Copyright Annmarie Geniusz. Used with permission.

7. Attaching warp to weaving frame. © Copyright Annmarie Geniusz. Used with permission.

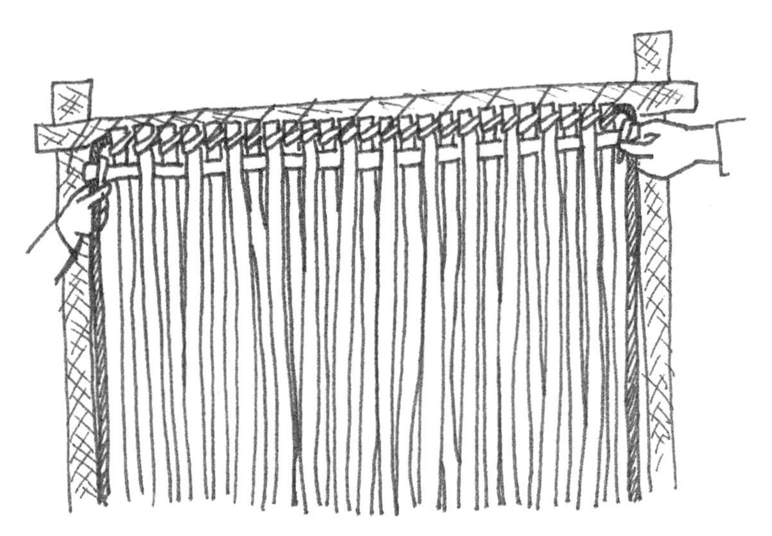

8. Weaving first weft and attaching both ends. © Copyright Annmarie Geniusz. Used with permission.

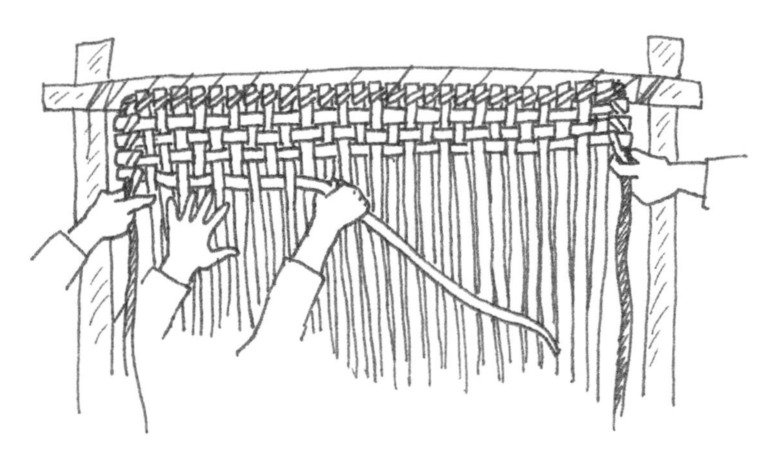

9. Continuing weaving pattern. © Copyright Annmarie Geniusz. Used with permission.

10. Pulling basswood cord across bottom of weaving. © Copyright Annmarie Geniusz. Used with permission.

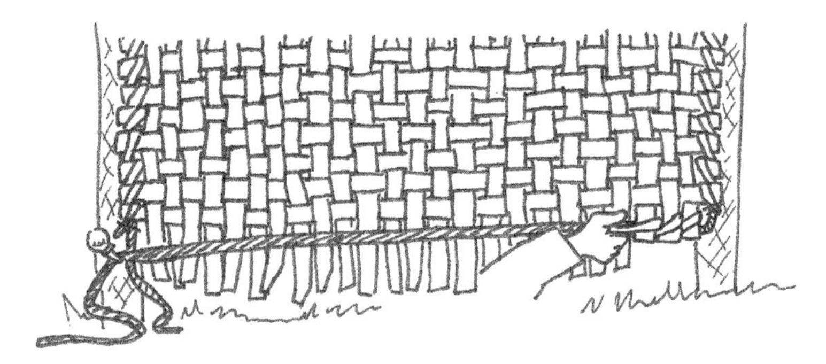

11. Finishing bottom of weft with same folding technique used to finish side and top edges. © Copyright Annmarie Geniusz. Used with permission.

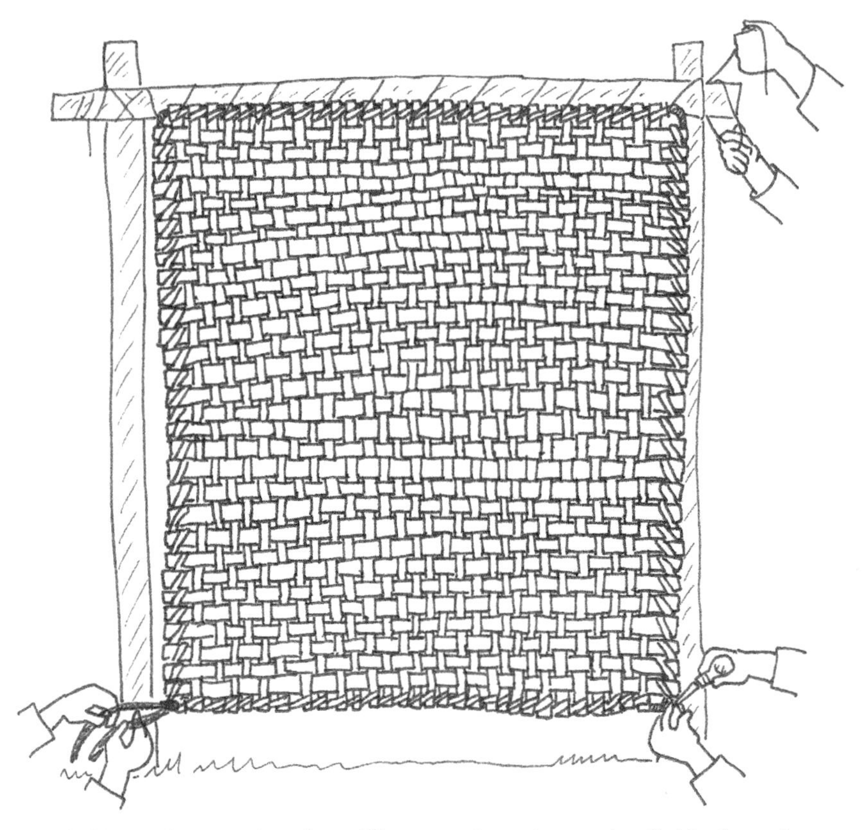

12. Tucking in loose ends, tying off loose cords, and removing finished mat from weaving frame. © Copyright Annmarie Geniusz. Used with permission.

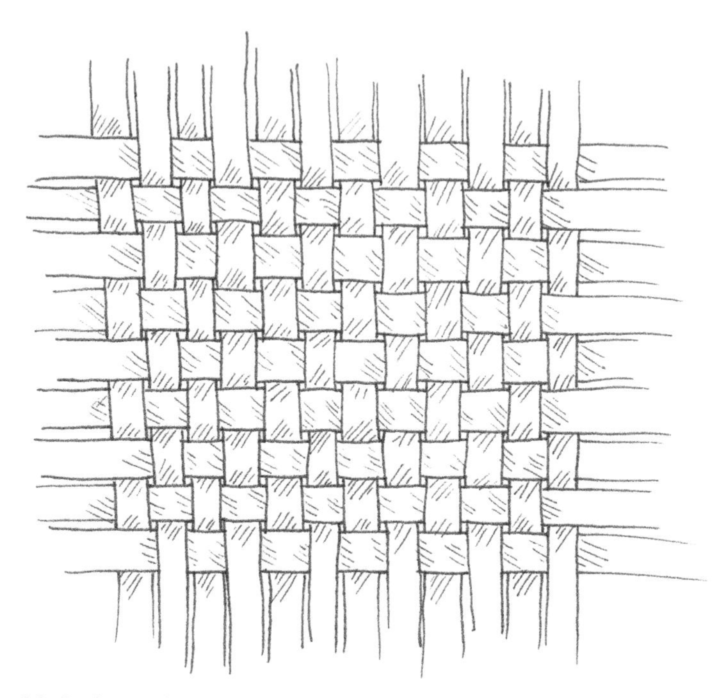

13. Basic weaving pattern. © Copyright Annmarie Geniusz. Used with permission.

14. Variation on basic weaving pattern. © Copyright Annmarie Geniusz. Used with permission.

## DIRECTIONS FOR GATHERING WIIGWAAS

The Anishinaabeg use wiigwaas (birch bark) to make a variety of items including roofs of lodges and storage containers. Keewaydinoquay cautioned her students to only gather wiigwaas from dead trees because she felt that most people would not know how to gather wiigwaas carefully enough so as not to kill the tree. She recommended using pieces of wiigwaas that had fallen off of rotting logs in the woods (1990b). Wiigwaas can be gathered from a living tree without killing the tree if one knows how to do it properly. Fresh wiigwaas is gathered in the spring when the sap has already begun to flow through the tree and the leaves have already begun to come out. George McGeshick, Sr., who has been gathering wiigwaas for decades on the border between Wisconsin and Michigan's Upper Peninsula, says that for his area, the time to gather wiigwaas is around the middle of June (pers. comm.). Angeline Williams says the best time to gather wiigwaas is when it rains because that is when it "rips off" well (EWV, notebook 19a). During the summer of 2004, George McGeshick taught some members of the Chicaugon Chippewa, my father, and myself how to peel wiigwaas in preparation for making a wiigwaasi-jiimaan (a birch bark canoe). He says that, contrary to what researchers such as Densmore and Huron Smith claim, it is not necessary to cut down a wiigwaasi-mitig before gathering wiigwaas to make a wiigwaasi-jiimaan (McGeshick, pers. comm.; Densmore [1929] 1979, 150; H. Smith 1932, 414). If one is careful not to cut too deeply into the tree's inner bark when cutting the bark, one will not kill the wiigwaasi-mitig. Basically, one should not cut so deep into wiigwaasi-mitig that one can see the wood in the place from which the bark has been peeled. McGeshick says that a wiigwaasi-mitig that has been peeled without exposing its wood will continue to live, but the white bark of that tree will never grow back again; instead the bark will become black (pers. comm.). He showed us a tree that he had once taken a large piece of wiigwaas from to make a canoe. The tree was still alive, but the place from which McGeshick had taken the wiigwaas was rough and black. When directing our birch bark peeling, McGeshick had us make a cut straight down wiigwaasi-mitig with a curved knife. He recommends using a utility knife with a small hook on the end, similar to the knife used by modern roofers. He also had us make one cut around the tree at the top of this vertical cut and one cut around the tree at the bottom of this vertical cut. We needed to do this on a straight section of the trunk, so we had to stop at parts that branched out into branches or roots. Then we were to pry the wiigwaas loose on either side of the vertical cut with the hooked knife and with a putty knife. Once we loosened the portion around the cut, the rest of the wiigwaas would just peel right off by pulling. Unfortunately, these particular wiigwaasi-mitigoog were not ready to be peeled yet, although others several miles away were ready. McGeshick said that the place we were peeling might have been in too much shade, causing the sap in the wiigwaasi-mitig to take longer to travel up the trees and causing the trees to take longer to be ready to peel (pers. comm.). Like the bark of giizhikaatig, wiigwaas can be rolled and tied in bundles and kept for later use. McGeshick says that once rolled, wiigwaas can be used years later as long as it is not stored in a place where it will decay quickly. He recommends keeping wiigwaas in a cool place, such as a basement (pers. comm.). When one wants to use the wiigwaas, one simply heats it to make it flexible again. Unless it is fresh from the tree, wiigwaas will crack when one tries to bend it if it is not warmed (EWV, notebook 20; WPA 1936–40, envelope 8, 38). McGeshick says one can easily warm the wiigwaas by unrolling the wiigwaas and leaving it in the sun to warm (pers. comm.). Peter Halfday, who worked for the WPA project on the Bad River Reservation, says that leaving wiigwaas in the summer sun can warm it, but he also suggests warming wiigwaas over a fire (WPA 1936–40, envelope 8, 38). Mary Geniusz often uses a clothes iron to heat up wiigwaas to make it pliable again, and she adds that, although many people try to soak wiigwaas to make it pliable, heat and not water is needed to get the resins in the wiigwaas active enough to make it workable (pers. comm.).[13] I use a warm-mist electric humidifier to bend wiigwaas.

### DIRECTIONS FOR MAKING AN APABIZON, THE BIRCH BARK TRAY INSIDE THE WAAPIJIPIZON

A baby is strapped into a waapijipizon (moss bag) and then tied into the dikinaagan (cradleboard). Inside the waapijipizon is the apabizon, a tray made of wiigwaas, which holds the aasaakamig (*Sphagnum* L., the moss used to diaper the baby). Mary Geniusz learned how to make one of these trays while she was working on her Masters of Indigenous Thought at Seven Generations Education Institute in Ontario, and I watched her make one during the summer of 2003. These instructions come from my observations while watching her make this tray.

First, take two pieces of wiigwaas and face them toward each other, so that the white outer barks are on the inside and the inner bark is on the outside. This position

13. I have heard from other people that soaking wiigwaas will not rejuvenate it, but I should add that my aunt, Sue Leather, who used to make containers out of wiigwaas, said that she would soak very thin pieces of wiigwaas in her bathtub for very long periods of time, and eventually they would become soft enough to bend into a makak. I have two of these containers, and the wiigwaas from which they are made is very, very soft and thin.

keeps the tray stiff. Then bend the bottom of this tray to form a sort of little shelf at one end, leaving the top straight. Geniusz bent the wiigwaas while it was still fresh off the tree, and she held it in place with two clothespins. Finally, sew the two pieces of wiigwaas together using wiigob, a fiber made from the inner bark of the basswood tree. This is the same fiber used to sew up a makak. To make wiigob, one peels bark off *wiigobaatig* (the basswood tree), and separates the outer and inner barks. The inner bark is then used right away for sewing or is dried and later soaked so that it becomes pliable again and is then used (Rose, pers. comm.).

Geniusz says that the dried aasaakamig is wrapped in a piece of clean cloth and put on this tray. The baby is laid on the tray and then tied into his or her moss bag. The baby is placed on this tray only when very young, before the fontanelle closes. The dikinaagan is not used before the fontanelle closes. When the tray is no longer used, the baby is simply tied into the moss bag, which is tied to the dikinaagan (pers. comm.).

### DIRECTIONS FOR STORING ITEMS IN A CONTAINER MADE OF WIIGWAAS

The Anishinaabeg use wiigwaas to make containers and to store certain items. Katherine Osogwin gives the following instructions for storing "dried pounded meat" in a container made of wiigwaas. Once the dried meat is put into the container, tallow is poured over the meat to seal it. Another piece of wiigwaas is placed over the top of the container, and this piece of wiigwaas is sealed with pine pitch. Osogwin adds that maple sugar is sealed and stored the same way. She says that the pine pitch is boiled in clear water and strained. When the water is cold, the pitch rises and is skimmed off the surface of the water. Then the pitch is boiled a second time until the water is out of it and it is the right consistency. Once this is done, fish oil or tallow is added to the pitch so that when the pitch is used to seal meat, "it won't crack or be sticky." The pitch is poured on top of the meat while it is warm to seal the meat, and at this time the pitch is transparent. She adds that this same pitch is used to make a pail made out of wiigwaas watertight so that it can be used for maple sugaring (EWV, notebook 20).

### DIRECTIONS FOR MAKING A WIIGWAASIBAK, A COVERING FOR A LODGE

A wiigwaasibak, or wiigwaasabakway, is a birch bark covering used on a waaginogaan and other shelters. Rose says she makes a wiigwaasibak by sewing pieces of wiigwaas together, end to end, with *wadab* (spruce root). The amount of wiigwaas used to make each wiigwaasibak depends on the size of the pieces of wiigwaas. Once all the pieces are sewn together, the two exposed edges of the wiigwaasibak are covered with a strip of cedar bark, which is sewn on with wadab, to stop them from splitting. A piece of wiigwaas, cut into a semicircle, is sewn onto the middle of this piece of cedar bark. The brown part of the wiigwaas, the underside of the bark, faces up, and the outer white bark is sandwiched against the cedar bark strip. Strings of wadab are attached to this semicircle of wiigwaas, and these strings are used to tie the wiigwaasibak to the wiigiwaam or waaginogaan (Rose, pers. comm.). The wiigwaasibak is tied onto the lodge with the white part facing upward. Rose adds that when describing this covering being tied to the lodge, one says, *"wiigwaasibak apakwen"* (put up the covering). When asking someone to help put up these coverings, one says, *"Apakwedaa"* (let's put up the coverings) (pers. comm.).

## DIRECTIONS FOR MAKING TORCHES

Wiigwaasi-mitig and giizhikaatig are also two trees used to make torches. Joe Wilson, of the Bad River Reservation, told Peter Marksman, who worked on the WPA Indian Research Project, that the Anishinaabeg make torches out of "cedar bark, birch bark, hazel brush, or basswood timber." To make a torch out of cedar bark, Wilson says, one crushes the bark into a "sheet of loose fibers," and then rolls this sheet tightly and ties it with strips of wiigob. One dips the roll into a mixture of melted pitch and powdered charcoal, which adds "firmness and durability" to the torch. The torch is about "the size of a man's forearm," and it is stored until needed. Before using the torch, the Anishinaabeg dip it in pitch once more. Wilson says that the Anishinaabeg make torches of hazel brush and basswood in a similar manner. He adds that the Anishinaabeg use blocks of cedar, which are split into thin boards, as reflectors and "as shields to conceal the occupants of the canoe" (WPA 1936–40, envelope 8, 37).

Peter Halfday, who worked for the same project on the Bad River Reservation, describes how to make a torch out of wiigwaas. It is assumed, given that Halfday mentions carrying this torch, that the torch described here is a waaswaagan (a hand-held torch). First, Halfday says, cut the wiigwaas into oblong sheets, and then hold each sheet over "a blazing fire until it begins to curl up." Then roll these sheets of wiigwaas into a torch. Halfday explains this process by saying, "The heating and the rolling is really a combination process. Beginning at the corner of the birch sheet, it is rolled towards a diagonal corner." The resulting torch is about three feet long and has a three-inch diameter. The torch looks like a "funnel," and is rolled very tightly. Halfday says that a torch rolled compactly like this "burns with a steady light," but he warns that a torch rolled too loosely "will flare up and burn out quickly." The Anishinaabeg wind wiigob around the thin lower end of the torch to carry it easily and safely (WPA 1936–40, envelope 8, 38).

# Glossary

2. PRONUNCIATION KEY FOR DOUBLE VOWEL OJIBWE

| *Letter or Symbol* | *Ojibwe (translation)* | *Roughly Equivalent Sound in English* |
|---|---|---|
| *Long Vowels* | | |
| aa | n*aa*nan (five) | f*a*ther |
| ii | n*ii*win (four) | kn*ee* |
| oo | b*oo*zh*oo* (hello) | b*oo*t, b*oa*t |
| e | b*e*zhig (one) | w*ei*gh |
| *Short Vowels* | | |
| a | niizht*a*n*a* (twenty) | c*u*p |
| i | n*i*swi (three) | it |
| o | ning*o*dwaaswi (six) | g*o* |
| *Consonants (all the rest sound as they do in English)* | | |
| g | nin*g*odwaaswi (six) | *g*oat |
| zh | niiz*h* (two) | trea*s*ure |
| *Other Sounds* | | |
| ' [glottal stop] | a'aw (that one) | o-oh! |
| nh [indicates nasalization of previous vowel] | giigoonh (fish) | |

*Note on pronunciation:* This orthography is made up of long and short vowels, the former of which are emphasized.

*Note:* The entries in this glossary are arranged in the following manner: **Head word** *part of speech:* definition (source).[1] The head words are Ojibwe words found throughout this text. Nouns are followed by a hyphen and a set of letters indicating their plural form. For example: **amik, -wag** : *amik* is a beaver, and *amikwag* is beavers. The parts of speech are abbreviated as follows:[2]

*na:* animate noun
*ni:* inanimate noun
*pc:* particle
*vai:* animate intransitive verb
*vai + o:* animate intransitive verb that can take an object
*vii:* inanimate intransitive verb
*vta:* animate transitive verb
*vta:* inanimate transitive verb

Parts of speech follow the head word. One or more definition is given for each head word, and the source for each word is given in parenthesis at the end of each entry.

**aadizookaan, -ag** *na:* the spirit of or character in a traditional or sacred story or a legend (Johnson, pers. comm.).
**aadizookaan, -an** *ni:* a traditional legend; all of one's relations; ceremonies ("Anishinaabe Wordlist" 2003).
**aaniin** *pc:* hello! greetings! (Nichols and Nyholm 1995, 18).
**aasaakamig, -oon** *ni:* moss used to diaper babies (Rose, pers. comm.; Nichols and Nyholm 1995, 18). Geniusz writes that Keewaydinoquay identified this moss as *Sphagnum* spp. (M. Geniusz 2005, 245).
**abinoojiiyens, -ag** *na:* baby (Nichols and Nyholm 1995, 3).
**ajijaak, -wag** *na:* sandhill crane (Nichols and Nyholm 1995, 7).
**amik, -wag** *na:* beaver (Nichols and Nyholm 1995, 8).
**amikwiish, -an** *ni:* beaver lodge (Nichols and Nyholm 1995, 8).
**animikii, -g** *na:* thunderbird, thunderer (Nichols and Nyholm 1995, 9).
**Anishinaabe, -g** *na:* an Indian, an Ojibwe person (Nichols and Nyholm 1995, 10).

1. Double Vowel Orthography was devised by Charles Fiero. For more detailed information, see Nichols and Nyholm 1995.

2. For further explanation on the abbreviation used in the glossary, see Nichols and Nyholm 1995.

**anishinaabe-gikendaasowin** *ni:* anishinaabe knowledge: not just information but also the synthesis of our personal teachings. Explicit form. ("Anishinaabe Wordlist" 2003).

**anishinaabe-inaadiziwin** *ni:* anishinaabe way of being, behavior, psychology. Explicit form. ("Anishinaabe Wordlist" 2003).

**anishinaabe-izhitwaawin** *ni:* anishinaabe culture, teachings, customs, history. Explicit form. ("Anishinaabe Wordlist" 2003).

**Anishinaabemowin** *ni:* Ojibwe language, language as a way of life ("Anishinaabe Wordlist" 2003).

**anishinaabewaki** *ni:* anishinaabe country. anishinaabewakiing: in anishinaabe country (Nichols and Nyholm 1995, 10).

**anishinaabe-wiinzowin, -an** *ni:* Indian name. Explicit form (Dennis Jones, pers. comm.).

**apabizon, -an** *ni:* the birch bark tray inside of the waapijipizon, or moss bag (Rose, pers. comm.).

**apakwe** *vai:* to put a roof on (something) (Rose, pers. comm.; Nichols and Nyholm 1995, 12).

**asemaa** *na:* tobacco ((Nichols and Nyholm 1995, 13). Geniusz says that Keewaydinoquay identified asemaa as *Nicotina rustica,* which is the same genus, but not the same species, as the plant used in the production of commercial cigarettes (M. Geniusz 2005, 15).

**ataaso** *vai + o:* store something (Nichols and Nyholm 1995, 14).

**ataasoowigamig, -oon** *ni:* a peaked house used to store foods (Nichols and Nyholm 1995, 14). Wheeler-Voegelin writes this word in singular form only: "ataso·gamik" (notebook 20).

**awesiinh, -yag** *na:* wild animal (Nichols and Nyholm 1995, 15).

**Baambiitaa-binesi** *na:* "Rhythm-beater, pace-setter (Proto-flicker)." Mentioned in the aadizookaan about the creation of giizhikaatig. Etymology uncertain. Keewaydinoquay writes this word: "Bahmbetah-Benaysee" (1977, 1).

**bibigwan, -an** *ni:* flute (Nichols and Nyholm 1995, 31).

**bineshiinh, -yag** *na:* bird (Nichols and Nyholm 1995, 33; "Anishinaabe Wordlist" 2003).

**Binesi, -wag** *na:* spirit bird ("Anishinaabe Wordlist" 2003).

**Biskaabiiyang** *vai:* an approach to research that attempts to decolonize the Anishinaabeg and anishinaabe-gikendaasowin. The stem verb here is *biskaabii,* a *vai* meaning "to return to oneself," also used in reference to decolonization ("Anishinaabe Wordlist" 2003; Horton, pers. comm.).

**biskaabii** *vai:* to return to oneself, also used in reference to decolonization ("Anishinaabe Wordlist" 2003).

**biskitenaagan, -an** *ni:* a container made of folded, rather than sewn, birch bark, often made in a hurry when one needs to collect sap or use as an emergency cooking dish. Also used for a cooking dish while traveling (Nichols and Nyholm 1995, 240; EWV, notebook 20). Wheeler-Voegelin writes this word: "skitenaganen."

**boozhoo** *pc:* hello! greetings! (Nichols and Nyholm 1995, 39).

**Chi-anishinaabe, -g** *na:* see Gichi-anishinaabe.

**dewe'igan, -ag** *na:* drum (Nichols and Nyholm 1995, 44).

**dibaajimowin, -an** *ni:* teaching, ordinary story, personal story, history story ("Ani- shinaabe Wordlist" 2003). Nichols and Nyholm have "story, narrative" (1995, 45).

**dikinaagan, -an** *ni:* cradle board (Nichols and Nyholm 1995, 162).

**doodooshaaboojiibik, -an** *ni:* dandelion, *Taraxacum officinale* (Rose, pers. comm.).

**gaagiizom** *vta:* to feast someone. Used when specifically stating which plant is being feasted (D. Jones, pers. comm.); to appease someone; apologize to someone (Nichols and Nyholm 1995, 50).

**gaa-izhi-zhawendaagoziyang** *vai:* that which was given to us in a loving way [by the spirits]; literally: "that which we are loved by the spirits" ("Anishinaabe Wordlist" 2003). Stem verb here is *zhawendaagozi,* a *vai* meaning "to be blessed, pitied, or fortunate" (Nichols and Nyholm 1995, 124).

**gaawiin wiiskiwizisii wii-maazhised** *sentence:* He or she is not living in balance; his or her behavior is not natural (i.e., it could lead to a death). Also **Gaawiin wiiskiwisiziiwag:** they are not living naturally, in balance. Etymology uncertain ("Anishinaabe Wordlist" 2003).

**gekek, -wag** *na:* hawk (Nichols and Nyholm 1995, 52).

**Gete-anishinaabe, -g** *na:* one of the old ones, an old-time Indian ("Anishinaabe Wordlist" 2003); also *Gichi-anishinaabe.*

**Gichi-anishinaabe, -g** *na:* an old-time Indian (Whipple, pers. comm.). **Gichi-manidoo** *na:* God, Great Spirit (Nichols and Nyholm 1995, 54). **giigoonh, -yag** *na:* fish (Nichols and Nyholm 1995, 58).

**gii-zhaaganashiiyaadizid** *vai:* one who is colonized; one who tries to live his or her life as a non-native, at the expense of being an Anishinaabe. Also said *gii-wemitigoozhiiyaadizid* ("Anishinaabe Wordlist" 2003). The verb stem here is *zhaaganashiiyaadizi,* a *vai* meaning "to be colonized."

**giizhik, -ag** *na:* white cedar (Nichols and Nyholm 1995, 61).

**giizhikaandag, -oog** *na:* cedar (Whipple, pers. comm.); cedar boughs (Nichols and Nyholm 1995, 61).

**giizhikaatig, -oog** *na:* northern white cedar, *Thuja occidentalis* L. (McGeshick, pers. comm.).

**gikendaasowin** *ni:* (anishinaabe) knowledge, not just information but also the synthesis of our personal teachings ("Anishinaabe Wordlist" 2003).

**Gimaamaanaan** *ni:* Our Mother, referring to Mother Earth (Rose, pers. comm.).

**gimishoomisinaanig** *na:* our grandfathers (Johnson 2006).

**gookooko'oo, -g** *na:* owl (Nichols and Nyholm 1995, 62).

**inaadiziwin** *ni:* anishinaabe way of being, behavior, psychology ("Anishinaabe Wordlist" 2003).

**inawendiwin** *ni:* interconnectedness, also our relationships within all of Creation ("Anishi-naabe Wordlist" 2003). Verb stem here is *inawendiwag,* a *vai* meaning "they are related to each other." Also, *inawem* is a *vta* meaning "to be related to someone" (Nichols and Nyholm 1995, 65).

**ishkaatig, -oog** *na:* a hardwood, sugar maple (McGeshick, pers. comm.).

**ishkodekaan, -an** *ni:* something used to transport fire from one place to another, a "lighter" (Nichols and Nyholm 1995, 69). Marie D. Livingston, who worked on the WPA project, refers to this as a "method of carrying fire," and she writes this word "ic-ko-de-kan" (WPA 1936–40, envelope 8, 39).

**izhitwaawin** *ni:* (anishinaabe) culture, teachings, customs, history ("Anishinaabe Wordlist" 2003).

**jiisakii** *vai:* to do the shaking tent ceremony; Nichols and Nyholm have "practice divination in a shaking tent" (1995, 245).

**madoodiswan, -an** *ni:* sweatlodge (Nichols and Nyholm 1995, 74).

**maji-aniibiish, -ag** *na:* poison ivy, *Rhus radicans* L. (Rose, pers. comm.).

**makak, -oon** *ni:* box, birch bark basket (Nichols and Nyholm 1995, 75). **makoons, -ag** *na:* bear cub. Diminutive of makwa (Nichols and Nyholm 1995, 76). **makwa, -g** *na:* bear (Nichols and Nyholm 1995, 76).

**makwa-miskomin, -an** *ni:* bearberry (specifically referring to the berries of this plant), *Arctostaphylos uva-ursi* L. Keewaydinoquay writes it: *Mukwa-Miskominan,* and translates it as "bear, his red berries" (1977, 10).

**manidoo, -g** *na:* spirit (Nichols and Nyholm 1995, 77).

**manoomin** *ni:* wild rice (Nichols and Nyholm 1995, 77). Note: the ending "min" suggests that this name does not refer to the entire plant.

**mashkiki, -wan** *ni:* medicine (Nichols and Nyholm 1995, 78).

**mashkikiiwikwe, -wag** *na:* medicine woman.

**mashkikiiwinini, -wag** *na:* medicine man.

**mashkosiw, -an** *ni:* grass, hay (Nichols and Nyholm 1995, 79).

**Midewaajimowin** *ni:* Midewiwin knowledge, what the "Midewiwin teach" ("Anishinaabe Wordlist" 2003).

**Midewiwin** *ni:* Mide, Midewiwin, Grand Medicine Society, a religious organization of the Anishinaabeg (Nichols and Nyholm 1995, 84).

**migizi, -wag** *na:* eagle (Nichols and Nyholm 1995, 84).

**miin, -an** *ni:* blueberry berry (Nichols and Nyholm 1995, 89).

**miinawaa** *pc:* and, also, again (Nichols and Nyholm 1995, 89).

**miinagaawanzh, -iig** *na:* blueberry plant (Nichols and Nyholm 1995, 89).

**mina'ig, -oog** *na:* "highland spruce" (Johnson 2006). White spruce, *Picea glauca* (Rose, pers. comm.).

**Mishi-makwa** *na:* The great bear. Mentioned in the aadizookaan about the creation of giizhikaatig.

**mishiikenh, -yag** *na:* snapping turtle (Nichols and Nyholm 1995, 275).

**mitig, -oog** *na:* tree (Nichols and Nyholm 1995, 88).

**mitigomin, -an** *ni:* acorn of the white oak tree, *Quercus alba* (Rose, pers. comm.).

**mitigomizh, -iig** *na:* white oak tree, *Quercus alba* (Rose, pers. comm.).

**mookijiwaninibiish** *ni:* spring water (McGeshick, pers. comm.).

**mukwa-miskomin:** see makwa-miskomin.

**name, -wag** *na:* sturgeon (Nichols and Nyholm 1995, 91).

**Nenabozho** *na:* a culture hero of the Anishinaabeg; many aadizookaanan are told about him. He is also called by other names, including Wenabozho in Minnesota. Often spelled "Nanabozho" in English. He is half-spirit and half-human, being the son of a manidoo who controls the West Wind and an Anishinaabe woman (Nichols and Nyholm 1995, 118).

**nigig, -wag** *na:* otter (Nichols and Nyholm 1995, 96).

**Niiyawen'enh, -yag** *na:* my namesake (Nichols and Nyholm 1995, 101).

**Nimishoomis wiigwaas** *na:* Grandfather birch, *Betula Papyrifera* Marsh (Keewaydinoquay 1990a). Keewaydinoquay says that this tree came to the Anishinaabeg much later than some of the other trees, so if one was intending to use a familial relationship name in reference to the birch, it would make sense to use "uncle," but she adds, "because of his tremendous importance and the terms grandfather and grandmother being terms of respect, he's called grandfather birch" (1988).

**Nookomis giizhik** *na:* Grandmother cedar, *Thuja occidentalis* L. (Keewaydinoquay 1988).

**ogiishkimanisii, -g** *na:* kingfisher (Nichols and Nyholm 1995, 105).

**Ojibwemowin** *ni:* the Ojibwe language (Nichols and Nyholm 1995, 105).

**omakakii, -g** *na:* frog (Nichols and Nyholm 1995, 106).

**omakakiibag:** frog leaf. "Mukikeebug (frog petal)" identified as "jewelweed" (Zichmanis and Hodgins 1982, 263), *Impatiens capensis* Meerb. and/or *Impatiens pallida*.[3]

**oshkaabewis, -ag** *na:* traditionally trained apprentice (M. Geniusz, pers. comm.); ceremonial attendant, ceremonial messenger (Nichols and Nyholm 1995, 110).

**ozhaaboomin, -an** *ni:* berry of gooseberry plant, *Ribes oxyacanthoides* (Rose, pers. comm.).

**ozhaaboominaganzh, -ag** *na:* gooseberry plant, *Ribes oxyacanthoides* (Rose, pers. comm.).

**waabooz, -oog** *na:* rabbit (Nichols and Nyholm 1995, 115).

**waaginigaan, -an** *ni:* domed lodge, wigwam (McGeshick, pers. comm.).

**waaginogaan, -an** *ni:* domed lodge, wigwam (Nichols and Nyholm 1995, 116).

**waapijipizon, -an** *ni:* moss bag, what baby is tied in before being tied into the dikinaagan (cradle board) (Rose, pers. comm.). Ron Geyshick has: *"wapigijibizun"* (1989, 22).

**waaswaagan, -an** *ni:* handheld torch (McGeshick, pers. comm.; Nichols and Nyholm 1995, 117).

**Waaswaaganing** *ni:* name for Lac du Flambeau Reservation in Wisconsin, which comes from the name for a handheld torch: *waaswaagan* (McGeshick, pers. comm.).

**waawaabiganoojiinh, -yag** *na:* mouse (Nichols and Nyholm 1995, 117).

**wadab, -iig** *na:* spruce root used for sewing (Rose, pers. comm.; Nichols and Nyholm 1995, 113).

**wawaazisii, -g** *na:* bullhead (Nichols and Nyholm 1995, 151).

**wazhashk, -wag** *na:* muskrat (Nichols and Nyholm 1995, 114).

**Wenabozho** *na:* a cultural hero of the Anishinaabeg; many aadizookaanan are told about him. He is also called by other names, including Nenabozho (Nichols and Nyholm 1995, 118).

**wiigiwaam, -an** *ni:* wigwam; lodge (Nichols and Nyholm 1995, 118). A shelter people live in that is shaped like a cone (Rose, pers. comm.).

**wiigob, -iin** *ni:* basswood fiber, the inner bark of the basswood tree used for sewing birch bark (Rose, pers. comm.; Nichols and Nyholm 1995, 119).

3. Zichmanis and Hodgins also refer to this plant by a French name, *Chou sauvage,* which they say they identified in *Flore Laurentine.* In this text, *"Chou sauvage"* is identified as *Impatiens capensis* Merrb (Marie-Victorin 1964, 399–400).

**wiigobaatig, -oog** *na:* basswood tree, *Tilia americana* (McGeshick, pers. comm.; Nichols and Nyholm 1995, 119).

**wiigwaas, -ag** *na:* birch tree (Nichols and Nyholm 1995, 119).

**wiigwaas, -an** *ni:* birch bark (Nichols and Nyholm 1995, 119; McGeshick, pers. comm.).

**wiigwaasaatig, -oog** *na:* birch tree, *Betula papyrifera* Marsh. (Rose, pers. comm.).

**wiigwaasabakway, -an** *ni:* birch bark covering; roll of birch bark roofing (Nichols and Nyholm 1995, 119; Whipple 2006b).

**wiigwaasibak, -oon:** *ni:* birch bark coverings for lodges (Rose, pers. comm.).

**wiigwaasi-jiimaan, -an** *ni:* birch bark canoe (Nichols and Nyholm 1995, 119).

**wiigwaasi-makakoon, -oon** *ni:* birch bark box, birch bark basket (Nichols and Nyholm 1995, 119).

**wiigwaasi-mitig, -oog** *na:* birch tree, *Betula papyrifera* Marsh. (McGeshick, pers. comm.; Nichols and Nyholm 1995, 119).

**wiikonge** *vai:* to give a feast (Nichols and Nyholm 1995, 119). Dennis Jones says this term is used generally, when one is giving a feast but not specifically saying who that feast is for (pers. comm.).

**wiinzowin, -an** *ni:* (Anishinaabe) name. Noun form of **wiinzo,** a *vai* meaning "have a name" (Nichols and Nyholm 1995, 120).

**wiiskiwizi** *vai:* to live in balance, to behave naturally ("Anishinaabe Wordlist" 2003). Etymology uncertain.

**zaka'aagan, -an ni:** headlight torch. Also said *saka'aagan* (McGeshick, pers. comm.).

**Zaka'aaganing ni:** "Place of the Headlight Torches," name for Mole Lake Reservation in Wisconsin. Also said: *Saka'aaganing.* (McGeshick, pers. comm.).

**zhaaganashiiyaadizi** *vai:* to be colonized, to try to live one's life as a non-native, at the expense of being an Anishinaabe; also said *wemitigoozhiiyaadizi* ("Anishina- abe Wordlist" 2003).

**zhawendaagozi** *vai:* to be blessed, pitied, or fortunate (Nichols and Nyholm 1995, 124).

**ziinzibaakwad** *ni:* maple sugar (Nichols and Nyholm 1995, 129).

# References

INTERVIEWEES

Richard Ford
Mary Geniusz
Laura Horton
Ken Johnson, Sr.
Dennis Jones
Peter Kaufman
George McGeshick, Sr.
John D. Nichols
Rose
Helen Hornbeck Tanner
Dora Dorothy Whipple

PRIMARY AND SECONDARY SOURCES

American Anthropological Association [AAA]. 1903. "The American Anthropological Association." *American Anthropologist* 5, no. 1: 178–90.

"Anishinaabe Wordlist." 2003. Collected by Mary Siisip Geniusz from panel of Ojibwe speakers at courses for the Seven Generations Education Institute Master of Indigenous Knowledge/Philosophy program, Ontario. For more information on this program, see www.7generations.org.

Archabal, Nina Marchetti. 1977. "Frances Densmore: Pioneer in the Study of American Indian Music." In *Women of Minnesota: Selected Biographical Essays,* ed. Barbara Stuhler and Gretchen Kreuter. St. Paul: Minnesota Historical Society Press.

———. 1979. Intro. to *Chippewa Customs,* by Frances Densmore. Reprint. St. Paul: Minnesota Historical Society Press.

Barrett, S. A. 1933. "In Memoriam." *Bulletin of the Public Museum of the City of Milwaukee* 7, no. 1: prelim.

Baraga, Frederic. [1878] 1992. *Dictionary of the Otchipwe Language.* Reprint, *A Dictionary of the Ojibway Language.* St. Paul: Minnesota Historical Society Press.

Benton-Banai, Edward. 1988. *The Mishomis Book: The Voice of the Ojibway.* St. Paul, Minn.: Red School House.

Blackbird, Andrew J. 1887. *History of the Ottawa and Chippewa Indians of Michigan: A Grammar of Their Language, and Personal and Family History of the Author.* Ypsilanti, Mich.: Ypsilantian Job Printing House.

Brockway, Lucile H. 1979. *Science and Colonial Expansion: The Role of the British Royal Botanic Gardens.* New York: Academic Press.

Carver, Jonathan. 1781. *Travels Through the Interior Parts of North America in the Years 1766, 1767, 1768. By J. Carver, Esq. Captain of a company of provincial troups during the late war with France.* 3rd ed. *To which is added, some account of the author, and a copious index.* London: Printed for C. Dilly, in the Poultry; H. Payne, in Pall-mall; and J. Phillips, in George-yard, Lombard Street.

Casagrande, Joseph B. 1955. "John Mink, Ojibwa Informant." *Wisconsin Archeologist,* n.s., 36, no. 4: 106–28.

Chamberlain, Alex F. 1900. "In Memoriam: Walter James Hoffman." *Journal of American Folklore* 13, no. 48: 44–46.

Cohoe, William. 1964. *A Cheyenne Sketchbook by Cohoe, with Commentary by E. Adamson Hoebel and Karen Daniels Petersen.* Norman: Univ. of Oklahoma Press.

Collins, Minta. 2000. *Medieval Herbals: The Illustrative Traditions.* London: British Library and Univ. of Toronto Press.

Cowan, C. Wesley. 1988. "Volney Hurt Jones, 1903–1982." *American Antiquity* 53, no. 3: 455–60.

Crockett, James Underwood. 1972. *Trees.* New York: Time-Life Books.

Davis, E. Wade. 1995. "Ethnobotany: An Old Practice, A New Discipline." In *Ethnobotany: Evolution of a Discipline,* ed. Richard Evans Schultes and Siri von Reis, 40–51. Portland, Ore.: Dioscorides Press.

Davidson-Hunt, Ian J., Phyllis Jack, Edward Mandamin, and Brennan Wapioke. 2005. "Iskatewizaagegan (Shoal Lake) Plant Knowledge: An Anishinaabe (Ojibway) Ethnobotany of Northwestern Ontario." *Journal of Ethnobiology* 25, no. 2: 189–227.

Deloria, Vine, Jr. 2001a. "American Indian Metaphysics." In Deloria and Wildcat 2001, 1–6.

———. 2001b. "Power and Place Equal Personality." In Deloria and Wildcat 2001, 21–28.

Deloria, Vine, Jr., and Daniel R. Wildcat. 2001. *Power and Place: Indian Education in America*. Golden, Colo.: American Indian Graduate Center and Fulcrum Resources.

Densmore, Frances. 1907. "An Ojibwa Prayer Ceremony." *American Anthropologist*, n.s., 9, no. 2: 443–44.

———. 1910. *Chippewa Music*. Bureau of American Ethnology Bulletin 45. Washington, D.C.: Government Printing Office.

———. 1913. *Chippewa Music II*. Bureau of American Ethnology Bulletin 53. Washington, D.C.: Government Printing Office.

———. [1928] 1974. "Uses of Plants by the Chippewa Indians." *Forty-fourth Annual Report of the Bureau of American Ethnology*, 1–274. Reprint, *How Indians Use Wild Plants for Food, Medicine and Crafts*. New York: Dover Publications.

———. [1929] 1979. *Chippewa Customs*. Bureau of American Ethnology Bulletin 86. Reprint, with introduction by Nina Marchetti Archabal. St. Paul: Minnesota Historical Society Press.

———. 1941. "The Study of Indian Music." *Annual Report of the Board of Regents of the Smithsonian Institution*, 527–50.

DiBella, Christina. 2004. *Volney H. Jones, 1903–1982*. Univ. of Michigan, Museum of Anthropology Web site. http://ww.1sa.umich.edu/umma/about/history/jones/ (accessed May 6, 2005).

Erdrich, Louise. 2003. *Books and Islands in Ojibwe Country*. Washington, D.C.: National Geographic.

Erichsen-Brown, Charlotte. 1979. *Medicinal and Other Uses of North American Plants: A Historical Survey with Special Reference to the Eastern Indian Tribes*. New York: Dover Publications.

EWV [Erminie Wheeler-Voegelin papers]. Box 30. Edward Ayer Manuscript Collection. Newberry Library, Chicago. EWV's notebooks are in folders 270 and 271 of box 30. The element list is in folders 267–69 of box 30. The ethno-linguistic survey is in folder 274 of box 30.

Ewen, Elizabeth, and Stuart Ewen. 2006. *Typecasting: On the Arts and Sciences of Human Inequality*. New York: Seven Stories Press.

Fanon, Frantz. 1968. *The Wretched of the Earth*. Trans. Constance Farrington. New York: Grove Press. Original work published in 1963.

Fierst, John T. 1986. "John Tanner: Troubled Years at Sault Ste. Marie." *Minnesota History* 50, no. 1: 23–36.

———. 1996. "Strange Eloquence: Another Look at *The Captivity and Adventures of John Tanner*." In *Reading Beyond Words: Contexts in Native History*, ed. Jennifer S. H. Brown and Elizabeth Vibert, 220–41. Peterborough, Ont.: Broadview Press.

Ford, Richard I. 1978. "Published Works of Volney H. Jones." In *The Nature and Status of Ethnobotany*, ed. Richard I. Ford, 419–28. Anthropological Papers 67. Ann Arbor: Museum of Anthropology, Univ. of Michigan.

———. 1998. Foreword to *Puhpohwee for the People: A Narrative Account of Some Uses of Fungi among the Ahnishinaabeg*. 2nd ed. DeKalb: Northern Illinois Univ. Press.

Foster, Steven, and James A. Duke. 2000. *A Field Guide to Medicinal Plants and Herbs of Eastern and Central North America*. Peterson Guide Series. 2nd ed. Boston: Houghton Mifflin.

Gelb, I. J. 1952. *A Study of Writing*. Rev. ed. Chicago: Univ. of Chicago Press.

Geniusz, Mary Siisip. 2005. "Blessings and Balance, Balance and Blessings, For from Balance Comes All Blessings, Ahow: The Anishinaabe Ethnobotany taught by Keewaydinoquay to Mary Siisip Geniusz, Oshkaabewis to Keewaydinoquay." Manuscript in the private collection of Mary Geniusz.

Geniusz, Wendy. 2005. "Keewaydinoquay: Anishinaabe-mashkikiikwe and Ethnobotanist." In *Papers of the Thirty-sixth Algonquian Conference*, ed. H. C. Wolfart, 187–206. Winnipeg: Univ. of Manitoba.

Gerard, John. 1597. *The Herball or General Historie of Plantes gathered by John Gerarde*. London: John Norton.

Geyshick, Ron. 1989. *Stories by an Ojibway Healer: Te Bwe Win: Truth*. With Judith Doyle. Toronto: Impulse Editions, Summer Hill Press.

Gilmore, Melvin R. 1932a. "The Ethnobotanical Laboratory at the University of Michigan." *Occasional Contributions from the Museum of Anthropology of the University of Michigan*, no. 1. Ann Arbor: Univ. of Michigan Press.

———. 1932b. "Importance of Ethnobotanical Investigation." *American Anthropologist*, n.s., 34, no. 2: 320–27.

———. 1933. "Some Chippewa Uses of Plants." *Papers of the Michigan Academy of Science, Arts, and Letters* 17:119–43.

Griffin, James B. 1978. "Volney Hurt Jones, Ethnobotanist: An Appreciation." In *The Nature and Status of Ethnobotany*, ed. Richard I. Ford, 3–19. Anthropological Papers 67. Ann Arbor: Museum of Anthropology, Univ. of Michigan.

Harshberger, John W. 1896. "The Purposes of Ethno-botany." *Botanical Gazette* 21, no. 3: 146–54.

———. 1928. *The Life and Work of John W. Harshberger, Ph.D.: An Autobiography*. Philadelphia: No publisher given.

Herron, Scott M. 2002. "Ethnobotany of the Anishinaabek Northern Great Lakes Indians." Ph.D. diss., Southern Illinois Univ. at Carbondale.

Hilger, M. Inez. 1951. "Menomini Child Life." *Journal de la Société des Américanistes de Paris* 40:163–71.

———. [1951] 1992. *Chippewa Child Life and Its Cultural Background.* Bureau of American Ethnology Bulletin 146. Reprint, with introduction by Jean M. O'Brien. St. Paul: Minnesota Historical Society Press.

———. 1952. *Arapaho Child Life and Its Cultural Background.* Bureau of American Ethnology Bulletin 148. Washington, D.C.: Government Printing Office.

———. 1957. *Araucanian Child Life and Its Cultural Background.* Smithsonian Miscellaneous Collections vol. 133. Washington, D.C.: Smithsonian Institution Press.

———. 1960. *Field Guide to the Ethnological Study of Child Life.* Behavior Science Field Guides vol. 1. New Haven, Conn.: Human Relations Area Files Press.

Hinsley, Curtis M., Jr. 1981. *Savages and Scientists: The Smithsonian Institution and the Development of American Anthropology, 1846–1910.* Washington, D.C.: Smithsonian Institution Press.

Hoffman, Walter James. 1888. "Pictography and Shamanistic Rites of the Ojibwa." *American Anthropologist* 1, no. 3: 209–30.

———. 1889. "Notes on Ojibwa Folk-Lore." *American Anthropologist* 2, no. 3: 215–24.

———. 1891. "The Midē'wiwin or 'Grand Medicine Society' of the Ojibwa." *Seventh Annual Report of the Bureau of Ethnology, 1885–1886,* 143–300. Washington, D.C.: Government Printing Office.

James, Edwin, ed. [1830] 1956. *A Narrative of the Captivity and Adventures of John Tanner (U.S. Interpreteur at the Saut. de Ste. Marie) During Thirty Years Residence among the Indians in the Interior of North America.* Reprint. Minneapolis: Ross and Hanes.

Johnson, Ken, Sr. 2006. In *Asemaa: Tobacco.* CD-ROM. Created by Wendy Makoons Geniusz and Annmarie Geniusz. In author's possession.

Johnston, Basil. 1976. *Ojibway Heritage.* Lincoln: Univ. of Nebraska Press.

———. 1995. *The Manitous: The Spiritual World of the Ojibway.* New York: HarperPerennial.

Jones, Volney H. 1935. "A Chippewa Method of Manufacturing Wooden Brooms." *Papers of the Michigan Academy of Science, Arts, and Letters* 20:23–30.

———. 1936. "Some Chippewa and Ottawa Uses of Sweet Grass." *Papers of the Michigan Academy of Science, Arts, and Letters* 11:21–31.

———. 1941. "The Nature and Status of Ethnobotany." *Chronica Botanica* 6, no. 10: 219–21.

———. 1946. "Notes on the Manufacture of Cedar-bark Mats by the Chippewa Indians." *Papers of the Michigan Academy of Science, Arts, and Letters* 32:341–63.

———. 1965. "The Bark of the Bittersweet Vine as an Emergency Food among the Indians of the Western Great Lakes Region." *Michigan Archaeologist* 11, no. 3–4: 170–80.

Judd, Neil M. 1967. *The Bureau of American Ethnology: A Partial History.* Norman: Univ. of Oklahoma Press.

Keewaydinoquay. 1977. *MukwahMiskomin or Kinnickinnick: "Gift of Bear": An Original Tale Never Before Recorded How to Use Bearberry for Teas, Emergency Food, Treating Diabetes and Internal Infections, Making Non-narcotic Smoking Mixtures.* n.p.: Miniss Kitigan Drum.

———. 1978. *Min: Anishinaabeg Ogimaawi-minan: Blueberry: First Fruit of the People.* n.p.: Miniss Kitigan Drum.

———. 1985. Audiotape, side A, of class lectures on mullein and cattails presented at the University of Wisconsin-Milwaukee, Oct. 3 and 8. In author's possession.

———. 1986. "Gifts that Grandmother Cedar Shares with Us as Told by: Keewaydinoquay." Videotape. *Native American Philosophy and Relationships to Plant Life.* Produced by Keewaydinoquay and Three Loons. Copyright Univ. of Wisconsin-Milwaukee Board of Regents.

———. 1988. Audiotape of class lectures on white cedar, balsam fir, and pines presented at the Univ. of Wisconsin-Milwaukee, Jan. 21. In author's possession.

———. 1989a. Videotapes of lectures presented at the Ethnobotany Workshop of the Miniss Kitigan Drum, Jan. 20–22. In author's possession.

———. 1989b. Videotapes of lectures presented at the Philosophy Workshop of the Miniss Kitigan Drum, Jan. 27–29. In author's possession.

———. 1990a. Videotape of lectures presented at the Philosophy Workshop of the Miniss Kitigan Drum. In author's possession.

———. 1990b. Videotape of lectures presented at the Herbal Workshop of the Miniss Kitigan Drum, Mar. In author's possession.

———. 1991a. Videotape of lectures presented at the Ethnobotany Workshop of the Miniss Kitigan Drum, Spring. In author's possession.

———. 1991b. Videotape of lectures presented at the Ethnobotany Workshop of the Miniss Kitigan Drum, Nov. In author's possession.

———. 1998. *Puhpohwee for the People: A Narrative Account of Some Uses of Fungi among the Ahnishinaabeg.* 2nd ed. DeKalb: Northern Illinois Univ. Press.

———. (n.d.*a*). Audiotape of class lecture on slides of various flowering plants and joe pye presented at the University of Wisconsin-Milwaukee. In author's possession.

———. (n.d.*b*). *The Cedar Story as Told by Keewaydinoquay.* Audio recording. In author's possession.

———. (n.d.*c*). Résumé. Personal papers. Copy in possession of Helen Hornbeck Tanner. On this résumé, Keewaydinoquay listed her name as "Keewaydinoquay, Margaret Peschell."

Kenny, Mary B. 2000. "Ojibway Plant Taxonomy at Lac Seul First Nation, Ontario, Canada." M.S. thesis, Lakehead Univ., Ontario.

Kinietz, Vernon. 1947. *Chippewa Village: The Story of Katikitegon.* Cranbrook Institute of Science Bulletin no. 25. Bloomfield Hills, Mich.: Cranbrook Institute of Science.

———. 1965. *The Indians of the Western Great Lakes, 1615–1760.* Ann Arbor: Univ. of Michigan Press.

Kinietz, Vernon, and Volney H. Jones. 1941. "Notes on the Manufacture of Rush Mats among the Chippewa." *Papers of the Michigan Academy of Science, Arts, and Letters* 27:525–37.

Laidlaw, G. E. 1922. "Ojibwa Myths and Tales." *Journal of American Folklore,* n.s., 1, no. 1: 28–38.

Lipsitz, George. 2006. *The Possessive Investment in Whiteness: How White People Profit from Identity Politics.* Rev. ed. Philadelphia: Temple Univ. Press.

Mallery, Garrick. 1886. "Pictographs of the North American Indians." *Fourth Annual Report of the Bureau of American Ethnology,* 3–256. Washington, D.C.: Government Printing Office.

———. 1893. "Picture Writing of the American Indians." *Tenth Annual Report of the Bureau of American Ethnology.* Washington, D.C.: Government Printing Office.

Marie-Victorin, Frère. 1964. *Flore laurentienne.* Montréal: Les Presses de l'université de Montréal.

Meeker, James E., Joan E. Elias, and John A. Heim. 1993a. *Plants Used by the Great Lakes Ojibwa.* Odanah, Wis.: Great Lakes Indian Fish and Wildlife Commission.

———. 1993b. *Plants Used by the Great Lakes Ojibwa.* Abridged version. Odanah, Wis.: Great Lakes Indian Fish and Wildlife Commission.

Memmi, Albert. 1965. *The Colonizer and the Colonized.* Trans. Howard Greenfeld. Boston: Beacon Press. Original work published in 1957.

Moerman, Daniel E. 1998. *Native American Ethnobotany.* Portland, Ore.: Timber Press.

Nichols, John D., and Earl Nyholm. 1995. *A Concise Dictionary of Minnesota Ojibwe*. Minneapolis: Univ. of Minnesota Press.

Niering, William A., and Nancy C. Olmstead. [1979] 2001. *National Audubon Society Field Guide to North American Wildflowers: Eastern Region*. Rev. ed. by John W. Thieret. New York: Chanticleer Press.

"Notes and News." 1937. "Obituary of Dr. Albert B. Reagan." *American Anthropologist*, n.s., 39, no. 1: 186–87.

O'Brien, Jean M. 1992. Intro. to *Chippewa Child Life and Its Cultural Background*, by M. Inez Hilger. Reprint. St. Paul: Minnesota Historical Society Press.

Peschel, Keewaydinoquay. [1978] 1998. *Puhpohwee for the People: A Narrative Account of Some Uses of Fungi among the Anishinaabeg*. Dekalb: Northern Illinois Univ. Press. Originally published by the Cambridge, Mass., Botanical Museum of Harvard Univ.

———. 2006. *Keewaydinoquay: Stories from My Youth*. Ed. Lee Boisvert. Ann Arbor: Univ. of Michigan Press.

Petersen, Karen Daniels. 1963. "Chippewa Mat-weaving Techniques." *Smithsonian Institution Bureau of American Ethnology Bulletin* 186, Anthropological Paper 67, 211–86.

Peterson, Lee Allen. 1977. *A Field Guide to Edible Wild Plants: Eastern and Central North America*. Boston: Houghton Mifflin.

Powell, John Wesley. 1881. Intro. to *First Annual Report of the Bureau of Ethnology to the Secretary of the Smithsonian Institution, 1879–1880*. Washington, D.C.: Government Printing Office.

———. 1891. Intro. to *Seventh Annual Report of the Bureau of Ethnology to the Secretary of the Smithsonian Institution, 1885–1886*. Washington, D.C.: Government Printing Office.

Prigge, Gene. 1981. "Community Ties Are Strong in Indian Life." *Milwaukee Journal Sunday Accent*, Feb. 22, 1, 7.

Radin, Paul. 1914. *Some Myths and Tales of the Ojibwa of Southeastern Ontario*. Geological Survey of Canada Memoir 48. Anthropological series no. 2. Ottawa: Government Printing.

———. 1924. "Ojibwa Ethnological Chit-chat." *American Anthropologist*, n.s., 26, no. 4: 491–530.

———. 1928. "Ethnological Notes on the Ojibwa of Southeastern Ontario." *American Anthropologist*, n.s., 30, no. 4: 659–68.

Reagan, Albert B. 1921. "Some Chippewa Medicinal Receipts." *American Anthropologist*, n.s., 23, no. 2: 246–49.

———. 1922. "Medicine Songs of George Farmer." *American Anthropologist,* n.s., 24, no. 3: 332–69.

———. 1924. "The Bois Fort Chippewa." *Wisconsin Archeologist,* n.s., 3, no. 4: 101–32.

———. 1927. "Picture Writings of the Chippewa Indians." *Wisconsin Archeologist,* n.s., 6, no. 3: 80–83.

———. 1928. "Plants Used by the Bois Fort Chippewa (Ojibwa) Indians of Minnesota." *Wisconsin Archeologist,* n.s., 7, no. 4: 230–48.

Rogers, Henry. 2005. *Writing Systems: A Linguistic Approach.* Malden, Mass.: Blackwell Publishing.

Rohde, Eleanour Sinclair. 1922. *The Old English Herbals.* New York: Longmans, Green.

Schoolcraft, Henry R. 1851–57. *Historical and Statistical Information Respecting the History, Conditions and Prospects of the Indian Tribes of the United States.* 6 vols. Philadelphia: Lippincott Grambo.

Schultes, Richard Evans, and Siri von Reis, eds. 1995. *Ethnobotany: Evolution of a Discipline.* Portland, Ore.: Dioscorides Press.

Smith, Huron H. 1923. "Ethnobotany of the Menomini Indians." *Bulletin of the Public Museum of the City of Milwaukee* 4, no. 1: 1–174.

———. 1928. "Ethnobotany of the Meskwaki Indians." *Bulletin of the Public Museum of the City of Milwaukee* 4, no. 2: 175–326.

———. 1932. "Ethnobotany of the Ojibwe Indians." *Bulletin of the Public Museum of the City of Milwaukee* 4, no. 3: 327–525.

———. 1933. "Ethnobotany of the Forest Potowatomi Indians." *Bulletin of the Public Museum of the City of Milwaukee* 7, no. 1: 1–230.

Smith, Linda Tuhiwai. 1999. *Decolonizing Methodologies: Research among Indigenous Peoples.* London: Zed Books.

Spencer, Robert F. 1978. "Sister M. Inez Hilger, O.S.B., 1891–1977." *American Anthropologist,* n.s., 80, no. 3: 650–53.

Speck, Frank G. 1915. *Myths and Folk-lore of the Timiskaming Algonquin and Timagami Ojibwa.* Geological Survey of Canada Memoir 71. Anthropological series no. 9. Ottawa: Government Printing.

Stowe, Gerald C. 1940. "Plants Used by the Chippewa." *Wisconsin Archeologist,* n.s., 21, no. 1: 8–13.

Takaki, Ronald. 1993. *A Different Mirror: A History of Multicultural America.* New York: Back Bay Books/Little, Brown and Co.

Tanner, Helen Hornbeck. 1991. "Erminie Wheeler-Voegelin (1903–1988), Founder of the American Society for Ethnohistory." *Ethnohistory* 38, no. 1: 58–72.

Treuer, Anton, ed. 2001. *Living Our Language: Ojibwe Tales and Oral Histories: A Bilingual Anthology.* St. Paul: Minnesota Historical Society Press.

USDA, NRCS. 2006. *The PLANTS Database.* http://plants.usda.gov. Data compiled from various sources by Mark W. Skinner. National Plant Data Center, Baton Rouge, LA 70874-4490.

Vastokas, Joan M. 1984. "Interpreting Birch Bark Scrolls." *Papers of the Fifteenth Algonquian Conference.* Ed. William Cowan, 425–44. Ottawa: Carleton Univ.

Vennum, Thomas, Jr. 1980. "Densmore, Frances Theresa, May 21, 1867– June 5, 1957. Ethnomusicologist." In *Notable American Women: The Modern Period: A Biographical Dictionary,* ed. Barbara Sicherman and Carol Hurd Green, 185–86. Cambridge, Mass.: Belknap Press of Harvard Univ. Press.

———. 1988. *Wild Rice and the Ojibway People.* St. Paul: Minnesota Historical Society Press.

Whipple, Dora Dorothy. 2006a. In *Asemaa: Tobacco.* CD-ROM. Created by Wendy Makoons Geniusz and Annmarie Geniusz. In author's possession.

———. 2006b. In *Izhi-ningo-biboon: One Whole Year.* CD-ROM. Created by Wendy Makoons Geniusz and Annmarie Geniusz. In author's possession.

Wisconsin Archeological Society [WAS]. 2003. *Wisconsin Archeological Society: Advancing Archaeology for 100 Years.* Wisconsin Archeological Society Web site. http://www.uwm.edu/Org/WAS/ (accessed Apr. 12, 2006).

WPA. 1936–40. "W.P.A. Indian Research Project Collection," directed by Sister Macaria Murphy. microfilm. Marquette Univ. Archives, Milwaukee, Wisconsin.

Zichmanis, Zile, and James Hodgins. 1982. *Flowers of the Wild: Ontario and the Great Lakes Region.* Toronto: Oxford Univ. Press.

Zinn, Howard. 2003. *A People's History of the United States: 1492–Present.* New York: HarperCollins.

# Index

Note: This is a decolonized index. It is hoped that it will serve as a model for future decolonized texts. As described in this book, the mechanisms of colonization have often bestowed the title of "expert" on researchers of indigenous knowledge rather than on the indigenous peoples themselves. One of the goals of decolonizing indigenous knowledge is to give proper credit to the people from whom these teachings came. Thus, in this index the names of researchers are listed but so are the names of their native consultants. Information given by each native consultant is listed under his or her name rather than under the name of the researcher, although cross-references are made. Another key aspect of decolonization is giving the proper respect to indigenous knowledge and to the origins of that knowledge as given by indigenous people. As explained in this book, Anishinaabe teachings tell us that many spirits, plants, and animals have shared information with us. Therefore, all of these beings are given separate entries in this index, often with subheadings describing the information they have given to us.

*Italic page numbers denote illustrations.*

Printed in the USA
CPSIA information can be obtained
at www.ICGtesting.com
CBHW031238230324
5612CB00002B/8

9 780815 638063